1980

THE
LUNAR
EFFECT

The unitary concept proposed here is compatible
with Einstein's unified field theory, with Darwin's
theory of evolution, and with Ludwig von Bertalanffy's general systems theory.

About the author

A. L. Lieber, M.D., is widely informed in the disciplines comprising the new science of cosmobiology. These disciplines include medicine, psychiatry, physics, biology, meteorology, astronomy, and biomagnetism. His discovery of lunar periodicity in human aggressive behavior was a historic moment.

Dr. Lieber is a psychiatrist in private practice in Miami, Florida, and on the faculty of the University of Miami School of Medicine. He is certified by the National Board of Medical Examiners and the American Board of Psychiatry and Neurology. From 1972 to 1975, he was psychiatric consultant to the City of Miami Police Department. He is the author of numerous scientific articles and a published poet.

About the producer

Jerome Agel's twenty book productions include 22 *Fires* (a novel); *The Cosmic Connection* and *Other Worlds* (with Carl Sagan); *Herman Kahnsciousness*; *The Making of Kubrick's 2001*; *The Radical Therapist*; *The Medium is the Massage* (with Marshall McLuhan); *I Seem To Be a Verb* (with Buckminster Fuller); *Fasting: The Ultimate Diet* and *Fasting as a Way of Life* (with Allan Cott, M.D.); *Understanding Understanding* (with Humphry Osmond, M.D.); and *A World Without—What Our Presidents Didn't Know*.

THE
LUNAR
EFFECT

*Biological Tides
and Human Emotions*

by Arnold L. Lieber, M.D.

produced by Jerome Agel

ANCHOR PRESS/DOUBLEDAY
GARDEN CITY, NEW YORK
1978

ISBN: 0-385-12897-5
Library of Congress Catalog Card Number 77-12870

To My Mentors

Marshall M. Lieber, M.D. (1904–74), my father and role model for open-minded scientific curiosity.

Seymour L. Alterman, M.D., who provided encouragement and the opportunity to initiate original clinical research and writing during my medical-student days.

James N. Sussex, M.D., whose courage, vision, and generosity enabled the work that comes to fruition in *The Lunar Effect*.

—A. L. LIEBER, M.D.

Contents

1. *The Enigmatic Presence*   1

For our ancestors, the Moon's power was beyond question. Donne's phrase "no man is an island" applies to all phenomena in the Universe. There is a very real physiological basis for some degree of transformation of a human being's appearance during full moon. My research into lunar rhythms in human aggression helps to explain why the werewolf legend has had such remarkable staying power in the imagination of the human race. New York's demonic "Son of Sam" killed on eight different nights—five of them were during new or full moon. In the year ending April 1977, there were nine full-moon suicides from the Golden Gate Bridge. Moon-driven acts of violence cannot be excused.

2. *The Moon and Murder*   15

Police and fire departments are convinced of the relationship between the Moon and violence. The graph of homicides in Dade County shows a striking correlation with the lunar-phase cycle. Life has biological high and low tides governed by the Moon. A build-up of body water can overload the system and actually alter personality. There is clear prejudice against work showing the Moon's influence on our daily life. Sun-sided temporal orientation overlooks lunar influences by ignoring lunar timing.

Organisms are either in a state of positive or negative receptivity relative to their natural environment. The moon-phase cycle is the key lunar cycle in the timing of plant, animal, and human activities. All hell broke loose when the Moon was only 217,000 miles from the Earth. The outburst of violence in the first three weeks of January 1974 was accurately predicted.

Man is a symphony of rhythms and cycles. The human reproductive system follows lunar time. Environmental forces that derive ultimately from cosmic cycles are encountered at birth and thereupon set the individual's biological rhythms.

All life forms are cosmic resonators. Mental illness was accompanied by a change in the electrical field state of the patient. Organisms are probably capable of perceiving changes in the Earth's electromagnetic field that are brought about by the movement of the Moon in relation to the Earth and the Sun— a sort of extrasensory perception. The pull of the Moon might trigger a faster heartbeat in an individual already under a strain. Because we now live exclusively by solar timing, we have been ignoring potentially lifesaving information.

The Moon distorts the Earth as if it were a rubber ball. The "Piccardi effect" on water suggests the possibility of an effect on body processes by cosmic events through the medium of electromagnetic fields. Only now are we starting to realize the Russians have a great deal of accumulated knowledge about the geophysical environment. The Moon may be responsible for disappearances in the Bermuda Triangle. There is evidence that living organisms, during the course of their evolution, have used electromagnetic fields to obtain information about changes in the natural environment. Electrophotography of the fingers of psychic healers shows dramatic changes when the healers "turn on" their healing power.

7. *The Moon in Evolution* 91

Evidence of the Moon's influence on evolution constitutes a major argument in favor of the theory of biological tides. Our precursors who took warning at full moon—and remained restless instead of sleeping—lived longer and left more progeny; this lunar pattern is ingrained in our genetic heritage. Electromagnetic fluctuations are received directly by the nervous system and somehow affect the mutation of chromosomes. When man is able to accept the "wolfish" tendencies within himself, he may no longer require a scapegoat—human or animal. Myths of the Moon and the Sun parallel the biological structure and function of thought. In overbalancing our emphasis on solar rationalism, we have slowed the evolution of creative thought.

8. *The Moon and Civilization* 99

If the earthquakes predicted for the next decade are accompanied by a massive behavioral upheaval throughout the population, we will have an impressive validation of some of the tenets of the ancient astrologers and, of course, of our own theories. A universal and unresolved problem in the study of natural environmental influences on human behavior is the basic unsuitability of statistics. A high price is paid in a world where only rational behavior is socially accepted. We must treat our unconscious minds and talents with respect if we wish to keep our heads together. The current desperation of our society can be seen most clearly by moonlight. It is the repression of the Moon's influence and what it stands for that brings about social tension, disharmony, and lamentable, often bizarre, results. It is a killer moon for individuals who are not psychically balanced or for a society too rigid to roll with the cosmic punch. Society's failure to come to terms with the intuitive side of human nature has resulted in a vast market for cosmic pseudoknowledge.

9. *The Biological Tides Theory* 109

We live in an electromagnetic world. My biological tides hypothesis states that the human body is susceptible to the same cosmic influences as is the Earth and that body processes ebb and flow with gravitational and electromagnetic tides. It is reasonable to assume that gravity exerts a direct effect on the water mass of the body, as it does on the water mass of the planet. The feeling that "this just ain't my day" may be sometimes extraterrestrial in origin. The finding of excessive bleeding tendencies in humans at the time of new and full moon provides further supporting evidence for biological tides. One of the noteworthy effects of the environment on consciousness is to reduce consciousness of the environment. A new field of science is evolving: cosmobiology, a unifying border science that will clarify the relationship of man to his universe according to the laws of nature as we know them or as we are finding out about them.

10. *Applied Lunar Knowledge*   125

Cosmic stresses may present an advantage to us because they are predictable. A dangerous biological high tide is treatable by addition of a simple chemical to the body fluid. If something we have learned can help to reduce the level of violence attributable to the influence of the Moon, we have made a worthwhile advance. There is much less violent crime in societies using lunar calendars. Older age groups in general are more susceptible to perturbations in the geophysical environment than are younger people. Manic depressives are exquisitely sensitive to sudden imbalances in the geomagnetic environment. With proper timing and a bit of caution, we can learn to live in harmony with cosmic forces and with our natural environment.

Introduction

There are recurring patterns in life whose meanings remain beyond us or at the horizon of our understanding. Forces and balances in our world manifest themselves in seemingly coincidental effects having no discernible connections. The influence of the Moon on our lives has been sensed for thousands of years, but only now has scientific reason been able to fathom the basic aspects of this influence.

Much of the evidence in this book is new. Information from a variety of disciplines that shed light upon the subtle effects of the natural environment on human life is presented here. The task has been to find the common patterns in this information. The *real* relationships are only beginning to emerge.

The Moon weighs 81,000,000,000,000,000,000,000 tons and circles the Earth at an average distance of 238,857 miles. There are still some people who believe it has absolutely no effect on our lives.

In the course of *The Lunar Effect: Biological Tides and Human Emotions,* you will meet some interesting people making some fascinating discoveries. The scope of investigation ranges from the observation of laboratory animals to the collection of world-wide weather data. The odyssey takes us from the depths of outer space to the still-being-explored workings of human biological systems. We are peeling away realms of the unknown.

The Moon has been omnipresent in man's environment. In league with the Sun, it has been a potent force in evolution. It indeed influences patterns of behavior. It helps to set the timing of the natural cycles by which we live. That "big chunk of green cheese" is an important link between man and the cosmos.

What does it all mean to you and me? Is it true, as Shakespeare wrote in *King Henry IV, Part 1*, that

> ". . . the fortune of us that are but moon's men
> doth ebb and flow like the sea . . ."

My own research "unearthed" data revealing lunar influence on violent behavior, with implications for psychiatry, medicine, and the behavioral sciences. There are also implications for normal daily life. People will think differently about themselves after reading this book.

Do you think of your skin as a boundary between you and the Universe? In many respects, it is not. To many cosmic forces, we present absolutely no barrier; in effect, we are transparent to them. *The Lunar Effect* is not concerned with our separateness from the world but, rather, our continuity with it.

Arthur Koestler reminds us that one of the cofounders* of the new universe—the astronomer and lawmaker of nature Johannes Kepler—wrote a number of serious treatises on astrology warning certain theologians, physicians, and philosophers "that, while justly rejecting the stargazers' superstitions, they should not throw out the child with the bath water. 'For nothing exists nor happens in the visible sky that is not sensed in some *hidden* manner by the faculties of Earth and Nature . . .' "

The first readers of this book have described it as a "now it can be told" tale. It certainly is a detective story of scientific discovery. The material that unfolds here *is* startling. But bear in mind, if you will, Sherlock Holmes's appropriate exhortation, "When you have eliminated the impossible, whatever remains—however improbable—must be the truth."

* The other founder, Tycho de Brahe, sent regrets that a forthcoming opposition of Mars and Jupiter, to be followed by a lunar eclipse, would keep him from welcoming personally the visiting Johannes Kepler. Brahe was just too busy—in February 1600.

I believe that you, too, will be convinced of the remarkable, even uncanny, power of the Moon.

In the original research that resulted in the biological tides theory, I was assisted by Carolyn R. Sherin, Ph.D.

Numerous scientists and technicians gave of their time, talent, and resources during the voyage of discovery. I wish to acknowledge with gratitude the following:

Joseph H. Davis, M.D., Dade County Medical Examiner; Samuel Gerber, M.D., and Lester Adelson, M.D., Cuyahoga County coroners; the late Milton Helprin, M.D., Medical Examiner of the City of New York; E. Wilson Purdy, Dade County Director of Public Safety; Thomas Carpenter, meteorologist, National Oceanic and Atmospheric Administration; Douglas Duke, astronomer, University of Miami; Fred Haddock, astronomer, University of Michigan; Richard Sherin, computer programmer, and Mitchell Manin, M.D., research assistant.

My research was supported in part by a grant from the United Way of Dade County, Florida.

In the research for, and the writing of *The Lunar Effect*, I was assisted by Jerome Agel, who approached me with the idea for this book, and by Fred Lazarus.

I am fascinated by Colin Wilson's account of the publishing history of Robert Graves's book on the Moon, *The White Goddess:*

> When the book was finished, odd things continued to happen. The first publisher who rejected it died of heart failure shortly afterwards. A second rejected it with a rude letter saying he could not make head or tail of it and he doubted whether anybody else could either; he dressed himself in women's underwear and hanged himself on a tree in his garden. On the other hand, says Graves, the publisher who accepted it— T. S. Eliot—not only got his money back, but also received the Order of Merit that year.

For those who are interested in a more intensive study of the phenomena described in *The Lunar Effect*, a bibliography appears at the end of the text.

A. L. Lieber, M.D.

Miami
March 1978

There is a tide in the affairs of men . . .
                    —*Julius Caesar*, IV, iii

I have no doubt the Moon has an effect on
human behavior.   —Carl Sagan

In the racial memory, the Moon is bad news, and
the lunatic is simply a fellow who happens to
remember better.   —*The New Yorker*

THE
LUNAR
EFFECT

1

The Enigmatic Presence

Five hundred million years ago, the Moon summoned
life out of its first home, the sea, and led it onto the
empty land. For as it drew the tides across the barren
continents of primeval earth, their daily rhythm exposed to sun and air the creatures of the shallows.
Most perished—but some adapted to the new and hos-
tile environment. . . . No wonder that the drama of
a launch engages our emotions so deeply. The rising
rocket appeals to instincts older than reason; . . .

—ARTHUR C. CLARKE

For most, if not all, of the human experience, the Moon has
been regarded as a mysterious and powerful entity. It has been
worshiped and feared and consulted.

For our ancestors, the Moon's power was beyond question.
They lived close to nature and observed that rhythms of life
were in harmony with the changing phases of the lunar cycle.
Modern man, on the other hand, treats the Moon with disre-
spect. Many scientists view it as nothing more than a big gray
cinder in the sky. Adding injury to insult, we have even walked
all over it.

Our ancestors knew something, however.

The uncanny power attributed to our nearest celestial neighbor by prescientific sages can now be linked to scientific
findings.

We are now learning that our only natural satellite wields a
pervasive influence on emotions, health, safety, and sanity. On-

going studies expose the Moon's relationship to human violence and fertility. It is a lover's moon; it is a killer moon.

The Moon alters our body chemistry.

We are, in a very real sense, in the grip of the Moon.

Ancient wisdom concerning the Moon and its power can no longer be dismissed as superstition. We owe it to ourselves to take a closer look.

It is easier to reject a whole body of knowledge than it is to test the validity of its concepts, especially concepts that to many are merely "old wives' tales." The line between fact and fancy is hard to define, hard to detect. The results of recent investigation are astonishing, and current scientific dogma may be shown to be no more than contemporary myth.

Nearly a century before Isaac Newton formulated the laws of universal gravitation, the founder of modern astronomy, Johannes Kepler (1571–1630), asserted that the Moon controlled the tides of the ocean. But he could not explain how. He could not link the effect to the cause. Galileo, a premier astronomer of the day, rejected Kepler's theory as "astrological nonsense." Kepler was indeed an astrologer as well, casting horoscopes to help keep body and soul together.

In Kepler's time, few people could tell the difference between astrology and astronomy. Galileo could, of course, and his wholesale rejection of astrology led him to disbelieve the influence of the Moon. Having laid the foundations of modern dynamics—the science concerned with moving bodies—he believed tides were caused by irregularities in the motion of the Earth. Newton, born a year after Galileo's death, was to explain the precise effect of the Moon on tides. What was once considered astrology is a basic principle of astronomy today.

Donne's phrase "no man is an island" applies to all phenomena in the Universe. Everything is interconnected, although we are not always able to demonstrate the connection.

Feeling funny? Take a look at the Moon. It's probably full.
Feeling violent? Take a look at the Moon. It's probably full.
Can't sleep? Take a look at the Moon. It's probably full.

Some doctors schedule themselves to be available to handle a heavier patient load at the time of full moon. Dr. Allan Cott, a prominent psychiatrist in New York City, informs me that he and the staff at the hospital at which he works can only explain sudden erratic patient behavior as being "triggered" by the Moon. He has noted that telephone callers have a tendency to rudeness during full-moon periods.

"The nuts really start pestering me during full moon," a Massachusetts congressman, Gerry Studds, has told me. "When I ask my secretary what kind of telephone calls we had overnight, she invariably responds during full moon, 'Oh, just the usual crazy stuff.'"

Until now, there has been no scientific support for the view of medical staffs that patient behavior is exaggerated during full moon. But the personal, empirical observations of emergency-room and psychiatric-ward personnel emphatically suggest that the Moon plays a key role in the behavior of normal *and* mentally aberrant people.

People working in institutions for the mentally retarded, for alcoholics, and for drug addicts have shared Moon-connected experiences with me.

An ambulance driver on New York's Lower East Side says that the street is always more active on a night of full moon. "On those nights, there is a holy mess of violent crimes and accidents."

Vernon Fox, professor of criminology at Florida State University, encounters similar experiences. In his classes, police officers indicate their belief that crime rates go up during full moon. Their belief is the result of individual perceptions based on personal observations.

The Chief of Police Inspectors for the City and County of San Francisco asserts "that most people in public service with any experience to speak of are possessed with empirical knowledge that there is *a definite relationship* between the phases of the Moon and human behavior."

A justice of the peace in Kingsland, Georgia, claims he has detected a marriage pattern during the four years he's been the justice. "It gets heavier on the full of the moon."

Belief in the power of the Moon to influence human behavior is not a superstitious practice of silly people. It is the informed opinion of experienced professionals who work with the public.

The Moon's effect on life is not limited to strange and even freakish human behavior. Its effect is widespread, and some people have been quick to take advantage of its predictable influences. The Moon has been a boon to hunters and fishermen. Old-timers have always done their fishing according to phases of the Moon. Certain fish bite at full moon; other types, at the quarters. STRIPERS TO FOLLOW FULL MOON TO MONTAUK says the headline over a field and stream column in the New York *Daily News*.

It seems natural that sea creatures would be most affected by the Moon because ocean tides, which are, after all, lunar-caused, are an important force in their environment. I believe there is increased metabolic activity in certain kinds of fish during certain phases of the Moon; because of the metabolic increase, the fish burn more energy, which in turn makes them hungry. Naturally, they will bite more frequently when they are hungry, and at just about anything.

In the Miami area, where I live and practice, the shrimpers are out en masse every full moon. At this time, shrimp predictably rise to the surface to feed. Bridges are clogged with whole families using nets, and tons of shrimp are caught.

Animal studies have shown that physical activity, metabolism, aggression, and sexual behavior are dramatically increased in many species at new and full moon. Predators like the wolf are active at full moon, a fact well known to trappers.

There now exists a range of scientific literature dealing with the influence of solar and lunar cycles on the planting, growing, and harvesting of crops. In many parts of the world, America included, there are farmers who plant by phases of the Moon— they always have. This practice is taken seriously by agronomists as well as by farmers, both modern and traditional. Popular beliefs and scientific investigation are coming together. Farmers' almanacs include Moon-oriented planting and harvesting guides. C. R. Trowbridge, publisher of *The Old Farmer's Almanac* (established in 1792), writes to me that plants bearing above-ground crops are supposed to grow better

OUTDOOR PLANTING TABLE, 1978

The best time to plant flowers and vegetables that bear crops above the ground is during the LIGHT of the Moon; that is, between the day the Moon is new to the day it is full. Flowering bulbs and vegetables that bear crops below ground should be planted during the DARK of the moon; that is, from the day after it is full to the day before it is new again. These moon days for 1978 are given in the "Moon Favorable" columns below.

The three columns below give planting dates for the Weather Regions listed. Plant one week later for every 500 feet above sea level. Weather regions 5 and South 16 are practically frost free.

Above Ground Crops Marked(*) E means Early L means Late	Weather Regions 1, 6, 9, 10, North 13		Weather Regions 2, 3, 7, 11, South 13, 15		Weather Regions 4, 8, 12, 14, 16	
	Planting Dates	Moon Favorable	Planting Dates	Moon Favorable	Planting Dates	Moon Favorable
*Barley	5/15-6/21	6/6-21	15-4/7	3/15-23	2/15-3/7	2/15-21
*Beans (E)	5/7-6/21	5/7-21, 6/6-19	4/15-30	4/8-21	3/15-4/7	3/15-23
(L)	6/15-7/15	6/15-19, 7/6-15	7/1-21	7/6-18	8/7-30	8/7-17
Beets (E)	5/1-15	5/1-5	3/15-4/3	3/25-4/3	2/7-28	2/23-28
(L)	7/15-8/15	7/20-31	8/15-30	8/19-30	9/1-30	9/17-30
*Broccoli (E)	5/15-30	5/15-21	3/7-30	3/9-23	2/15-3/15	2/15-21
(L)	6/15-7/7	6/15-19	8/1-20	8/4-17	9/7-30	9/7-15
*Brussels Spr.	5/15-30	5/15-21	3/7-4/15	3/9-23	2/11-3/20	2/11-21
*Cabbage Pl.	5/15-30	5/15-21	3/7-4/15	3/9-23	2/11-3/20	2/11-21
Carrots (E)	5/15-30	5/23-40	3/7-31	3/25-31	2/15-3/7	2/23-3/7
(L)	6/15-7/21	6/21-7/4	7/7-30	7/20-30	8/1-9/7	8/9-31
*Cauliflower (E)	5/15-30	5/15-21	3/15-4/7	3/9-23	2/15-3/7	2/11-21
Plants (L)	6/15-7/21	7/6-18	7/1-8/7	7/6-18, 8/4-7	8/7-30	8/7-17
*Celery (E)	5/15-6/30	5/15-21, 6/6-19	3/7-30	3/9-23	2/15-28	2/15-21
(L)	7/15-8/15	8/4-15	8/15-9/7	9/3-7	9/15-30	9/15
*Corn, Sw. (E)	5/10-6/15	5/10-21, 6/6-15	4/1-15	4/8-15	3/15-30	3/15-23
(L)	6/15-30	6/15-19	7/7-21	7/7-18	8/7-30	8/7-17
*Cucumber	5/7-6/20	5/7-21, 6/6-19	4/7-5/15	4/8-21, 5/7-15	3/7-4/15	3/9-23, 4/8-15
*Eggplant Pl.	6/1-30	6/6-19	4/7-5/15	4/8-21, 5/7-15	3/7-4/15	3/9-23, 4/8-15
*Endive (E)	5/15-30	5/15-21	4/7-5/15	4/8-21, 5/7-15	2/15-3/20	2/15-21
(L)	6/7-30	6/7-19	7/15-8/15	8/4-15	8/15-9/7	9/3-7
*Flowers	5/7-6/21	5/7-21, 6/6-19	4/15-30	4/15-21	3/15-4/7	3/15-23
*Kale (E)	5/15-30	5/15-21	3/7-4/7	3/9-23	2/11-3/20	2/11-21
(L)	7/1-8/7	7/6-18	8/15-31	8/15-17	9/7-30	9/7-15
Leek Pl.	5/15-30	5/23-30	3/7-4/7	3/25-4/6	2/15-4/15	2/23-3/7
*Lettuce	5/15-6/30	5/15-21, 6/6-19	3/1-31	3/9-23	2/15-3/7	2/15-21
*Muskmelon	5/15-6/30	5/15-21, 6/6-19	4/15-5/1	4/15-21	3/15-4/7	3/15-23
Onion Pl.	5/15-6/7	5/23-6/4	3/1-31	3/1-7, 25-31	2/1-28	2/1-6, 23-28
*Parsley	5/15-30	5/15-21	3/1-31	3/9-23	2/20-3/15	3/9-15
Parsnip	4/1-30	4/1-6, 23-30	3/7-31	3/25-31	1/15-2/4	1/25-31
*Peas (E)	4/15-5/7	4/15-21	3/7-31	3/9-23	1/15-2/7	1/15-23
(L)	7/15-30	7/15-18	8/7-31	8/7-17	9/15-30	9/15
*Pepper Pl.	5/15-6/30	5/15-21, 6/6-19	4/1-30	4/8-21	3/1-20	3/9-20
Potato	5/1-15	5/1-5	4/1-15	4/1-4	2/10-28	2/23-28
*Pumpkin	5/15-30	5/15-21	4/23-5/15	5/7-15	3/7-20	3/9-20
Radish (E)	4/15-30	4/23-30	3/7-31	3/25-31	1/21-3/1	1/25-2/6
(L)	8/15-30	8/19-30	9/7-30	9/17-30	10/1-21	10/17-21
*Spinach (E)	5/15-30	5/15-21	3/15-4/20	3/15-23	2/7-3/15	2/8-21, 3/9-15
(L)	7/15-9/7	8/4-17, 9/1-7	8/1-9/15	8/4-17, 9/1-15	10/1-21	10/1-15
*Squash	5/15-6/15	5/15-21, 6/6-15	4/15-30	4/15-21	3/15-4/15	3/15-23
*Swiss Chard	5/1-30	5/7-21	3/15-4/15	3/15-23	2/7-3/15	2/8-21, 3/9-15
*Tomato Pl.	5/15-30	5/15-21	4/7-30	4/8-21	3/7-20	3/9-20
Turnip (E)	4/7-30	4/23-30	3/7-30	3/25-30	1/20-2/15	1/25-2/6
(L)	7/1-8/15	7/1-4, 20-31	8/1-20	8/1-2, 19-20	9/1-10/15	9/17-30
*Wheat, Winter	8/11-9/15	8/11-17, 9/3-15	9/15-10/20	10/3-15	10/15-12/7	10/1-13
Spring	4/7-30	4/8-21	3/1-20	3/9-20	2/15-28	2/15-21

with the increasing moon, and root crops do better if planted when the moon is waning.

Given examples of the Moon's influence in so many diverse ways on everyday life, we must ask ourselves: Is there any truth in the old wives' tales? My research on the Moon and human aggression has demonstrated there is substance to many of them.

The beliefs and traditions about the Moon's power over life do little to explain exactly *how* the various effects are brought about—such causal questions cannot yet be fully answered. Even the origin of our sister ship in the sky remains shrouded in mystery. Some scientists believe the Moon was once part of the Earth itself. According to this theory, the Moon (diameter: 2,160 miles; weight: 81,000,000,000,000,000,000 tons) was torn out of the surface of the Earth, leaving the great hole now filled with water and called the Pacific Ocean. The Moon is indeed of about the same density as the surface rock of the Earth's continents—the part that would have been torn away. But what cataclysmic force could have been responsible for such a world-rending event? Nobody has come forward with a sensible suggestion.

Another school of thought on the origin of the Moon—this one has as its standard bearer the heretical cosmogonist Immanuel Velikovsky—maintains that the Moon was *captured* by the Earth. But if the Moon was once a passing planet, where did it come from? How was it captured? Why didn't it hit the Earth?

However it got there—an average of 238,857 miles from the third planet from the Sun—the librational Moon influences Earth activity. How can we approach the question of discerning where and how the Moon enters into our life? Where best to start?

Moonlore is world wide. It has been told and retold throughout history. The mythographer and poet Robert Graves has written that the cult of the Moon goddess was at one time present throughout the entire European area and was only later replaced by patriarchal religions, first pagan, then Christian. Just how persistent Moon worship was can be seen by its survival, sometimes combined curiously with Christian worship. A historian "recorded that 'the wild Irish' still knelt before the new moon and recited the Lord's Prayer." Sometimes Moon worship was a startling throwback: ". . . in 1453 a butcher and a labourer of Standon, Hertfordshire, were formally accused of maintaining that there was no god save the sun and the moon." (To quote the historian Keith Thomas.)

An ordering of the natural world is implicit in such beliefs. The moonlore of India tells us that illness worsens on days of full moon and new moon, and those who are very ill are likely to die on those days. The Indians note that during certain phases of the Moon various species of insects and reptiles become poisonous. (This strange effect may be documented someday by animal studies.) The Indians also believe some medicinal herbs possess stronger healing qualities than usual during certain lunar phases. One particular herb taken during full moon is said to be effective against asthma. (In chapter 10, we shall see how modern psychiatry may be moving toward cosmic timing in the use of some medication.)

The connection between Moon and mental illness forms a significant chapter of moonlore. A study by Christian Medical College in Vellore, India, for instance, revealed that 58 per cent of the population were convinced the Moon influenced mental illness; at least 10 per cent would still seek out a witch to treat mental disorder. In Iceland, it is said that if a pregnant woman sits with her face turned toward the Moon, her child will be a *luna*tic. In Brazil, mothers hide their newborn children to prevent the moonlight from affecting them. A psychiatric study in the United States indicated that attacks of manic-depressive illness may be lunar-timed.

The power of the Moon often touches the darker side of the human soul. Although the Moon goddess was considered beneficent and the source of fertility, she was bloodthirsty as well. Anat, Moon goddess of Phoenician myth, often went berserk and killed for pleasure. For bloodthirsty deeds, none surpassed Kali Ma, dark Hindu moon goddess. Known as the black mother, she wore corpses for earrings, and human skulls formed her grisly necklace. In Greek myth, the Moon was not only the chaste goddess Diana but also the terror-striking Hecate, who ruled the storm and was the patroness of witches. We still celebrate Hecate's festival—Halloween.

Ancient societies, as their myths tell us plainly, knew there was a definite connection between the Moon and violence. This association is found around the world in the form of the werewolf legend. Wherever the wolf has roamed, people have told chilling tales of men and women who assumed the form of wolves and preyed upon cattle and human beings. Although stories of actual physical transformation are doubted by most people, it must be kept in mind that a legend so widespread must represent some important cultural and behavioral truth. In addition, there is a very real physiological basis for some degree of transformation of a human being's appearance during full moon (see chapter 7). Because it concerns violence in connection with the Moon, the werewolf legend provided an appropriate cultural background to my investigation of lunar periodicity in human aggressive behavior.

Most scientists have totally rejected werewolves as a fit subject for investigation because of prejudice against anything that deals with magical transformation. Some historians, psychologists, and anthropologists have dealt with the subject, realizing its importance for what it reveals about man's relation to his natural environment. From them we learn that lycanthropy —the fabled transformation of human into wolf—was considered a form of lunacy, a madness caused by the Moon. The Romans believed the weird metamorphosis might be brought on by angering the Moon goddess. It seems likely the metamorphosis theory originated in some form of Moon worship, which was probably universal in archaic society. In their rites, shamans and witch doctors always have taken on the form of totem animals. Lycanthropy probably originated in such a manner, the wolf being an important totem animal.

One fascinating theory, put forward by the psychologist Robert Eisler, speculates that the werewolf legend dates from the dim past, when human beings learned to hunt. It is fairly certain our ancestors were food-gathering vegetarians and that they did not become hunters until later in their development.

The change to hunting may have been brought about by a change in environment, such as the onset of an ice age. Plentiful vegetable food died out as the glaciers advanced, and hunting—bloody, vicious, violent—became necessary. Did early man react to this development as a traumatic event?

When men were forced by circumstance to adopt the practice of hunting—the way of the wolf—they found themselves in competition with this animal. Both man and wolf preyed upon herds of reindeer and developed a mutual respect for each other's hunting methods and abilities. It was inevitable that shamans would want to partake of the power of the wolf through ceremonies of imitative magic. These rites were intended to transform the shaman symbolically into a wolf in order to gain the valuable knowledge of that animal.

The relationship worked the other way as well. Wolves followed the camps of humans and scavenged their leavings. In this way, the dog was first domesticated from varieties of wolf.

The wolf became the emblem of the carnivorous way of life into which our vegetarian ancestors were plunged. Although the wolf is, in its relations with other wolves, one of the kindest and most loving of animals, it is to humankind the symbol of cruelty. It was in the company of the wolf, both competitive and co-operative (that is, hunting with dogs), that man fell from grace. This epic moment is ingrained in human memory.

It could not have escaped the notice of our ancestors that the wolf had a special relationship with the Moon. The baying of wolves at the Moon would have been a constant serenade to the humans following the wild herds. Under a bright full moon, wolves would communicate by howling to each other and then gather for the hunt. Moonlight provides a great advantage for the nocturnal marauder, and man was quick to emulate his fanged competitor. (A friend who lives up north informs me that his wolfish-looking dog will take off on moonlit nights to hunt and won't return until dawn—exhausted, but happy.)

Lycanthropy was a respected magical practice, linked to the vital activity of hunting. How did it come about that the werewolf became a terrifying creature, in popular opinion, an unnat-

ural monster? With another change in climate, the withdrawal
of the glaciers and the subsequent development of agriculture,
the importance of the hunter in the economy diminished. The
domestication of cattle followed that of dogs, which were then
employed to herd the cattle. Man's acquired taste for meat
could be satisfied without hunting. The wolf became a pest, a
brigand stealing livestock from the farmer and the nomad.

The werewolf lost his importance as a cultural symbol of the
means of survival. Robert Graves has written that the keepers
of domesticated animals in ancient Arcadia conceived of a
different role for the werewolf, more in line with the needs of
their way of life: "The Arcadian religious theory is that a man
is sent as an envoy to the wolves. He becomes a werewolf for
eight years, and persuades the wolf-packs to leave man's flocks
and children alone during that time." That role was eventually
forgotten. The wolf became the enemy of civilization, and the
werewolf was alienated from society. He is an archetypal monster, fearsome yet pathetic. The werewolf is a reminder of the
human race's fall from grace. He is the individual who re-
members too well, when he feels the pull of the full moon, the
blood trauma of the first hunters.

How real are werewolves? Some psychotic individuals, imag-
ining themselves to be wolves, have acted in a wild and aggressive manner, tearing at raw meat, raving, avoiding human contact, caring nothing for personal comfort or protection from
the elements. Reports of this kind of behavior are not un-
common. A case of lycanthropy has been described in the
American Journal of Psychiatry. At the full of the moon, a
forty-nine-year-old married woman would become psychotic
and believe she was a wolf—and would act like a wolf! During
these periods, she experienced strong homosexual urges, felt
sexually aroused and tormented, and had irresistible zoophilic
drives and masturbatory compulsions, which culminated in the
delusion of a wolflike metamorphosis. (The syndrome should
be regarded as a symptom complex, the investigators concluded, and not as a distinct diagnostic entity.)

In France, many tales are told of humans who, at the dark of
the moon, have the power to summon and lead packs of

wolves. It is well known to those who have studied wild-canine behavior that a human may be imprinted as the pack leader in the minds of wolves. Some people have raised packs of wolves.

In the case of the deranged person who imagines himself a werewolf, it is obvious there are other emotional components to his violent psychotic lapse besides sadism. This individual alienates himself from humanity and suffers great deprivation. He punishes himself for his personal guilt, whatever it may be. In a historical-cultural sense, this individual may be suffering for the collective guilt of the race.

The connection of the werewolf's activity with the Moon is traditional and easy to understand in light of the wolf's love of the Moon. But it continues to be a mysterious relationship. My research into lunar rhythms in human aggression throws light on this relationship. It helps to explain why the werewolf legend has had such remarkable staying power in the imagination of the human race.

Modern sadists have not been reluctant to exploit the strange power of the werewolf legend. In Germany, where the legend had been widespread, Nazi terrorists formed the Werewolf Organization in the 1920s. (Was it merely parenthetical that Hitler's nickname was "Wolf"?) At the end of the Second World War, the Werewolf Organization was supposed to continue, in guerrilla fashion, the struggle against the Allies who were occupying Germany. However, the members had enough trouble simply staying alive. Historical regression was a major theme of Nazi ideology, and the werewolf, with its regression to an archaic and violent way of life, was a fitting choice to symbolize the monstrosity of the fascist regime.

The werewolf is by no means the only criminal inspired by the Moon. Typical, and one of the best known of lunar anecdotes, is the case of Charles Hyde, the English laborer who was the model for Robert Louis Stevenson's Jekyll and Hyde character. Hyde was inspired by both the new moon and the full moon to commit criminal acts that would never have entered his mind "normally." Hyde contended in court that—by reason of lunacy—he was not responsible for his crimes. His defense of moon madness was of no avail. The court sentenced him to jail, in 1854. Then, as now, Moon-driven acts of violence cannot be excused.

Speculations on the Jekyll-Hyde theme are fascinating. Dr. E. A. Janino, of Lynn, Massachusetts, raised the possibility that both Jack the Ripper and the Boston Strangler were genuine lunatics, driven to their savagery by the Moon's influence. New York's demonic "Son of Sam" killed on eight different nights—five of them were during new or full moon. Sarah Moore shot at President Ford during a period of full moon. The Symbionese Liberation Army kidnaped Patricia Hearst during full moon. In the year ending April 1977, there were nine full-moon suicides from the Golden Gate Bridge. During full moon in August 1977, a sniper murdered six strangers in Hackettstown, New Jersey, then committed suicide. A man in Rockford, Illinois, is said to have butchered his six children during a full moon in January 1978.

Albert Einstein Hospital in New York City reports that neurotics do not seem strongly affected by the Moon. The effect seems to be reserved for those most truly deserving of the label "lunatic." Of course, normal people often observe the weird effects of the Moon.

The Moon has inspired prophets, philosophers, physicians, astronomers, and poets:

> The Moon has always exerted a magic and a madness
> to know more. The first observations of the Moon,
> said Flammarion, did not make less noise than the
> discovery of America; many saw in them another
> discovery of a new world much more interesting than
> America, as it was beyond Earth."
> —Timothy Harley

> . . . dearest man-in-the-moon
> I used to fear moonlight
> thinking her my mother.
> —Erica Jong

The Moon . . . is a face in its own right,
White as knuckle and terribly upset.
It drags the sea after it like a dark crime.
— Sylvia Plath

And when the clear moon, with its soothing influences,
rises full in my view,—from the wall-like rocks,
out of the damp underwood, the silvery forms of past
ages hover up to me, and soften the austere pleasure
of contemplation.

— Goethe's *Faust*

The precious things put forth by the Moon.
— Deuteronomy, 33:14

Richard Mead, a prominent eighteenth-century English physician, made these case-observations involving the influence of the Moon on the health of various persons:

> Doctor Pitcairne's case is remarkable . . . he was
> seized, at nine in the morning, the very hour of the
> new moon, with a sudden bleeding at the nose, after
> an uncommon faintness.
> That the fits of the asthma are frequently periodical,
> and under the influence of the Moon, and also of the
> weather . . .
> A more uncommon effect of this attractive power is
> related by the learned Kirchringus. He knew a young
> gentlewoman, whose beauty depended upon the lunar
> force, insomuch that at full moon she was plump and
> very handsome; but in the decrease of the planet so
> wan and ill favored, that she was ashamed to go
> abroad; till the return of the new moon gradually gave
> fullness to her face, and attraction to her charms.

Along with Dr. Mead, such prominent men as Robert Boyle, Francis Bacon, and Henry More believed in the biological influence of the Moon. Dr. Benjamin Rush, the father of American psychiatry, recognized and wrote of the relationship he found between moon phases and human maladies, especially mental disturbances. Physicians and natural philosophers throughout history—Heraclitus, Aristotle, Paracelsus, Maimonides, and many others—were aware of a relationship between the lunar cycle and physical and emotional health. Anaïs Nin once wrote: "In watching the moon she acquired the certainty of the expansion of time by depth of emotion, range and infinite multiplicity of experience."

What we know about the Moon is tantalizing but insignificant, compared with what we don't know about it. From the question of its origin to the question of how it exerts its power on our daily lives, the Moon remains enigmatic. When asking difficult questions, we must remember that meaningful connections are often not revealed until separate fields of knowledge co-evolve. The important role of cosmic rays as producers of mutations could not be understood until the fields of astrophysics and genetics had reached certain stages of development. Connective forces that appear so weak as to be inconsequential may prove to be very strong indeed, but in an unexpected way. Weak forces, for instance, often act as triggers for stronger ones. In addition, weak effects acting in a rhythmic, resonating manner may build to a considerable pitch. (Using this principle, the Hebrews rhythmically marched around the city of Jericho, creating a resonating effect that brought down the walls.)

Cosmic rhythms, often weak but regular and persistent through time, have not failed to leave their mark on our environment and on our evolution. We must not ignore the subtle effects of such forces: in many cases, they may prove to be the keys to understanding changes in the broader phenomena of life. Our understanding of the Moon's influence is proving to be such a key. The doors we seek to unlock open inward, on the *mind*, and outward, on the *cosmos*—vast realms of the unknown.

2

The Moon
and Murder

It is the very error of the moon;
She comes more near the earth than she was wont,
And makes men mad.
                              —Shakespeare's *Othello*

Shakespeare's insight into the motivation of human behavior
gives his characters a timelessness, assuring their survival in literature. Othello's irrational violence, linked to the "error of the
moon," reflects a common belief in the Bard's time and antici-
pates present interest in the role of the Moon in human aggression. Although there is now a considerable body of information
on the Moon's influence, most of it is subjective or based on
casual surveys that lack careful *scientific* methodology.

Police and fire departments have come to be convinced of
the relationship between the Moon and violence, having collected data that support their observations. The New York City
Bureau of Fire Investigation found that arson cases increased
up to 100 per cent at full moon. The Philadelphia police department noted a rise in arson and crimes against the person at
times of full moon. The Los Angeles and Miami police departments reported similar experiences. In Phoenix, there has been
an average of twenty-five to thirty more calls for the fire department on nights of full moon, according to Manuel Benitez, sen-
ior fire investigator.

These reports were not produced with rigorous methods and, therefore, do not constitute statistical proof. To get at "the truth," one must dig for hard facts that can be substantiated by statistical procedures. This calls for tough scientific detective work.

I became interested in the Moon's influence through the same sort of personal observations that have excited the interest of so many people. My involvement began when I was a medical student in training at Jackson Memorial Hospital in Miami. While working on the mental wards there, my curiosity was aroused by a peculiar pattern of events. I noticed there were recurring periods when behavior of patients was more disturbed than usual. These periods would occur seemingly for no apparent reason. They would last for a few days. The patients would then quiet down and the ward return to its normal routine. I couldn't help but wonder what caused the unexpected periods of commotion among the patients. Although circumstances prevented me from doing anything at the time to satisfy my curiosity, the question haunted me in the years that followed.

After a four-year tour as a flight surgeon in the U. S. Air Force, I decided, in 1969, to take a residency in psychiatry. Human behavior became my primary field of interest. When I returned to work in the mental wards, it was not long before I again noticed the unexplained periods of disturbed behavior. This time I was determined to discover the cause. Questioning the ward staff, I learned the nurses and attendants were familiar with these periodic occurrences. Almost jokingly, they invariably attributed the outbursts to the onset of full moon. Not knowing whether to believe them, I asked the staff of the emergency room about the phenomenon, and they too said they had noticed periods of restless and odd behavior among patients arriving for treatment during full moon. Police and rescue personnel bringing patients into the emergency room concurred.

One day, hesitantly, I brought up the subject in conversation with some of the medical residents. They told me there were periodic rashes of bleeding ulcers and epileptic seizures on the

medical wards and that the staffs there attributed then
influence of full moon.

I wasn't sure how seriously the speculation was being
Over a period of time I became convinced that hospital pe
nel—with many years' experience—really believed what
were saying. They had to! They had observed the coincidence
of freakish behavior and full moon too often to deny it.

My first effort at investigation took the form of a survey of
the psychiatric wards. Could social changes account for the periodic disturbances? I reasoned that changes of staff, the admission of particularly provocative patients, and even the weather
could be responsible for the disturbed behavior. However, none
of these could account for the *recurring regular pattern*. Could
it indeed be that the experienced workers I had questioned
were right? Was it, in fact, *full moon* that accounted for the
mysterious outbursts? If so, how?

To me, it seemed worth further investigation, even at the
risk of derision and ridicule from my colleagues. My attitude
became one of open-minded determination: "If it's bunk, let's
disprove it; if there is something to it, let's find out for certain." It seemed obvious to me that if the findings were positive, an important contribution would be made to health care.
For example, hospitals could increase the staff on the psychiatric ward at times of peak demand.

I thus began what developed into an extensive study of
moon phases as a possible factor in violent human behavior.

In the department of psychiatry at the University of Miami
Medical School, research projects had to be sanctioned by the
chairman. I knew a convincing research protocol would have to
be generated. The attitude of most scientists was that studies
associated with superstition, folklore, and mythology were a
foolish waste of time. I undertook an extensive review of the
scientific literature in order to learn what had been done on the
subject of possible lunar effects on human behavior. I turned
up several studies that had been published during the first half
of the twentieth century. The results of these studies were
mixed; some showed positive findings and just as many showed
negative results.

The strikingly even distribution of positive and negative results appeared to be due to a recurrent pattern in publication:
whenever a paper appeared showing a positive correlation between human behavior and the lunar cycle, another report
would quickly follow that repudiated the findings. This was my
first indication of what a scientific "hot potato" I was dealing
with. Critical reviews of many of the papers pointed out defects
in methodological approaches. Clearly, it would take ironclad
proofs to convince the skeptics. Which was fine with me.

I knew that in searching out lunar influences I would be
dealing with "small effects" rather than with gross, clear-cut
differences. This meant that I would have to collect a prodi-
gious amount of accurate data over a long period of time. If I
were to show meaningful correlations, careful selection of the
violent behavior to be measured was indispensable. In addition,
a way had to be found to convert calendar dates based on solar
time to a lunar time-frame. Prior studies had used crude techniques to accomplish this conversion. Fortunately, by the time
I began the study, in 1970, astronomical technology and modern computers made sophisticated and accurate conversion easy
to accomplish.

I searched for a variable by which I could measure violent
human behavior. I needed one that could be easily quantified
and treated as hard statistical data. To make sense of the information collected, I needed a simple, straightforward statistical
approach for assessing the findings.

While pondering these prerequisites, I came across an article
in the Miami *Herald* reporting a research project, by meteorologists at the University of Miami, that showed a lunar
influence on the formation of tropical storms and hurricanes.
The project had been conducted on a world-wide basis. Intrigued, I contacted one of the researchers, Ronald Holle, and
he introduced me to the world of meteorology. Meteorologists
have long-term accurate records on which to base their work.
They have long suspected the Moon of influencing the weather
and now have a simple procedure for measuring and assessing
local events according to lunar time. This procedure was subsequently modified and used by my research team. (I also learned

that meteorology had unquestionably documented an *atmospheric tide* that was reflected in precipitation levels and storm formation.)

One floor above the meteorological laboratories at the University of Miami was the department of physics. Mr. Holle introduced me to Dr. Douglas Duke, professor of physics and astronomy, who was immediately interested in my ideas on lunar effects on human behavior. He readily agreed to tutor me in the basics of modern astronomy. His suggestions, together with those of Mr. Holle and of Tom Carpenter of the National Oceanic and Atmospheric Administration, in Washington, were instrumental in helping me formulate an informed approach to my research.

I presented my proposal to Dr. James N. Sussex, chairman of the department of psychiatry at the University of Miami School of Medicine. He was aware of the prejudice against researching lunar influence on human behavior. There was even a possibility of risk to his own reputation, through association. He was, however, able to foresee the potential benefits of valid scientific findings. Not only was Dr. Sussex supportive and encouraging, he even advanced the seed money for the project from his own departmental research funds. Without his help my investigation would never have been able to get off the ground.

Thus, fortified in morale and funding, I began my research vigorously. I had to find a variable to measure emotional upheaval. If the Moon stood for romance, yet for violence, for sanguine thought, yet for emotional disturbance, how could such diverse effects be quantified? I was up against the variety and complexity of human behavior in general, not to mention considerable individual variability (which is why lunar anecdotes stand out). Human behavior is complex and confusing when viewed as a whole, but freakish events tend to stick in the memory.

We simply don't have adequate ways of quantifying human behavior. Experimental psychologists use a variety of rating scales to measure behavior. These scales, however, are subject to the prejudices of the experimenters who design them and of

the observers and interpreters who use them. I don't have
confidence in them. If an observer is looking for certain behaviors, he is more likely to find them. It is as simple as that. I was
determined from the outset to avoid observer biases.

I teamed with Dr. Carolyn Sherin, a clinical psychologist experienced in the application of statistical methods. After much
searching and discussion, Dr. Sherin and I chose homicide as
our variable for the measurement of violent behavior. Murder
is a violent act, and it can be, in most cases, easily pinpointed
in time. It has universally accepted medical and legal defini-
tions. We did not have to differentiate one type of homicide
from another, which would have been a source of bias. We
needed to know only that the act occurred and at what exact
time it occurred.

The matter of timing presented an important requirement
for the study: we had to learn the exact *time of injury* for the
murder. Time of death was *not* important. The victim might
die hours, days, or even weeks after the injury. Time of injury
denotes time of attack and therefore is the critical variable in
the precise timing of a violent act.

We were fortunate to be conducting our study in Miami. Dr.
Joseph H. Davis, the Dade County (Miami) Medical Examiner, has a national reputation for precise and meticulous conduct of his office. For his comprehensive records, he collects,
whenever possible, time of injury as well as time of death for
all violent deaths. In addition, his records are computerized
and ready for immediate use. This beautifully ordered data
sample was quite literally in our own back yard—Dr. Davis's
office is located in the rear of the hospital with which I am
affiliated.

I wish to re-emphasize that time of injury is of the utmost
importance. Time lag between injury and death as well as the
controversy over the exact definition of death were irrelevant to
our study. The violent emotional upheaval we wanted to measure is marked in time by attack and injury. This criterion also
kept our study simple and accurate.

Our requirement of knowing time of injury did not make our
work easy when it came to duplicating the study in another lo-

cation. In order to lend validity to findings we derived, we knew we had to replicate our study elsewhere. For this reason, we decided to use the homicide records of New York City.

The nation's largest city has about ten times as many murders as Miami, and one could reasonably assume its records would be thorough. Alas, one can't assume anything about New York. The city did not document time of injury. We had to search elsewhere for data as accurate as the Dade County records. We found them in Cuyahoga County (Cleveland), Ohio.

After eliminating homicides for which the exact time of injury was not known, we had 1,887 cases from Dade County spanning a fifteen-year period and 2,008 cases from Cuyahoga County covering thirteen years. The two samples were more than sufficient.

In looking for any influence of moon phases on human behavior, we assumed such effects would indeed be small. A great deal of accurate data would be needed to confirm any effect of the Moon. In my opinion, many studies failed to yield convincing results because they used samples that were too small or time periods that were too short.

We proceeded to convert the dates of the samples to lunar time. Conversion was absolutely necessary in order to avoid contamination by known social periodicities in human behavior. These periodicities—holidays, weekends, Saturday night hell-raising, blue Mondays, payday sprees, et cetera—are timed by our standard solar calendar. Had we organized our information on the solar calendar, all of these patterns would have shown up plain as day. But they would have masked the more subtle effects of our natural environment, especially the pure lunar rhythms. We needed to keep our information independent of solar time.

Meteorologists came to our rescue. They use a calendar called the lunar-synodic decimal scale. This lunar calendar doesn't, of course, concern itself with sunrise and sunset (or, in Buckminster Fuller's view, sunsee and sunclipse). By using a lunar calendar, we eliminated confusion.

The homicides were plotted on a graph according to the

lunar calendar. This provided us with a picture of the frequency of murder relative to the lunar cycle, but the chart was not used as statistical proof. For proof we resorted to a more rigorous statistical method. We counted the number of homicides occurring in the periods seventy-two, forty-eight, and twenty-four hours before and after each phase of the moon. The number of homicides in each of these "time windows" was compared with the number expected purely by chance.

The results were astounding! The graph of homicides in Dade County showed a striking correlation with the lunar-phase cycle. *The homicides peaked at full moon!* The graph

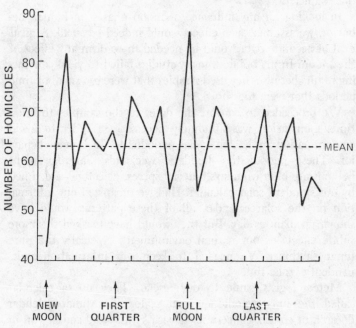

Homicides occurring in Dade County, Florida, over a fifteen-year period, plotted in relation to the lunar synodic cycle.

showed a trough leading to new moon, with a secondary peak immediately after new moon.

It has been known since Kepler's time that at new and full moon the pull of the Moon is strongest. Our results indicated that murders became more frequent with this increase in the Moon's gravitational force. The peaks on the graph were significantly greater than could be expected by chance. We had proved statistically there *was* a relationship between moon phases and murders. Although we had demonstrated a relationship, we had not established there was a causal effect.

Our results encouraged us to look further. When the homi-

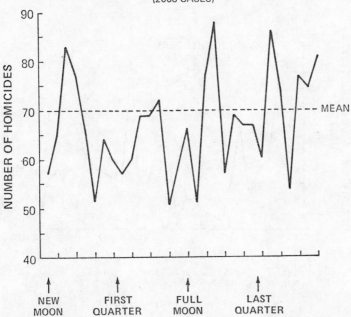

Homicides occurring in Cuyahoga County, Ohio, over a thirteen-year period, plotted in relation to the lunar synodic cycle.

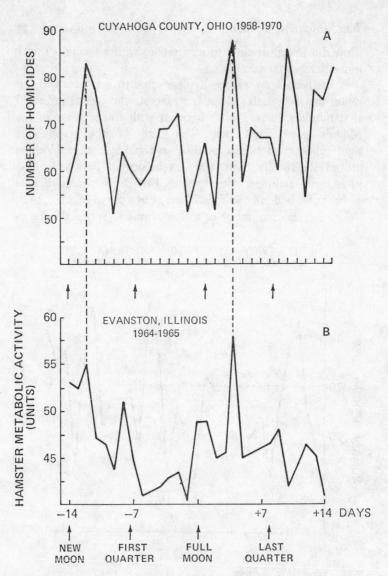

Comparison of the Cuyahoga County homicide curve (A) with Dr. Frank Brown's hamster activity curve (B) plotted in Evanston, Illinois. Cleveland is at latitude 41.30 N. and longitude 81.41 W.; Evanston is at latitude 42.02 N. and longitude 87.41 W. Note the striking coincidence of peaks relative to the lunar synodic cycle.

cides from Cuyahoga County were plotted on a graph, they showed a pattern similar to our Dade graph—but the peaks of the Cuyahoga graph were displaced to the right. That is, they occurred *later* than the Dade peaks. They did not coincide with full and new moon. They came about three days after both full and new moon. These shifted peaks approached statistical significance.

The mixed results of our studies posed difficult but intriguing questions. The Cuyahoga study fell short of confirming the striking pattern we had found in the analysis of Dade County records. But why did the graphs look similar? It seemed as if there were some sort of hidden relationship. Could the shifting of the Cuyahoga peaks to the right represent a lag in lunar effect due to the different geographical location?

We know that ocean tides, governed by the Moon, vary from one place to another. Not only does the timing of the tides differ locally, but the amplitude of the tides differs with respect to latitude. The difference between high tide and low tide around the equator is usually measured in inches, while the amplitude of tides occurring near the poles can be measured to the extent of fifty feet or more! We asked ourselves what effect latitude might have on the timing of lunar cycles.

Geographical variation in lunar effects was common in weather observations and animal studies. Animal experiments performed by Dr. Frank A. Brown, Jr., a pioneer in the field of biological rhythms, were of particular relevance to our enigmatic findings. Dr. Brown studies the effects of cosmic cycles on a variety of plant and animal life in his laboratory at Northwestern University, in Evanston, Illinois. He discovered a lunar periodicity in the metabolic activity of hamsters. Plotting his graph of hamster activity against the phases of the Moon showed practically the same lagged peaks we had found in the Cuyahoga County homicide chart! The hamsters were turning their treadmills in Evanston, which, in terms of latitude (42°N.), is a neighbor of Cleveland (41°/30′ N.). The correlation led us to suspect that the Moon exerts a similar effect on the activity of humans and animals at similar latitudes.

How could we explain these intriguing comparisons? How could the Moon possibly cause periodic changes in behavior?

In pursuit of the answers, I developed the theory of biological tides. This theory confronts the problem from two perspectives. The first approach attempts to explain the direct effect of the Moon's gravitational pull on living organisms. The second approach describes the Moon's indirect effect as mediated by the Earth's electromagnetic field.

The theory of biological tides, which is elaborated in chapter 9, is designed to make a contribution to a new understanding of the nature of man in relation to his natural environment. The solar system—and even the Universe—must now be considered as part of our natural environment.

In simple terms, the human being can be viewed as a microcosm made essentially of the same elements as the surface of the Earth, and in similar proportion. This is in harmony with ancient philosophical and religious themes in which the cosmos was seen to be reflected in man. We are finding provocative scientific parallels to historic ideas. Like the surface of the Earth, man is about 80 per cent water and 20 per cent solids. I believe the gravitational force of the Moon, acting in concert with the other major forces of the Universe, exerts an influence on the water in the human body—in you and in me—as it does on the oceans of the planet. Life has, I believe, biological high tides and low tides governed by the Moon. At new and full moon these tides are at their highest—and the Moon's effect on our behavior is its strongest.

Body water is contained in three main compartments. The *intravascular* water compartment is the water in the blood. It is of about the same chemical composition as sea water. *Extracellular* water floats free among the tissues, bathing the cells of the body. The water content of the cells themselves is the *intracellular* water. Water migrates freely among the three compartments. When we drink water, we replenish our body water. We eliminate water to keep the balance and prevent bloating. However, the process of maintaining fluid balance can be thrown off. Any number of shifts in body processes can cause an increase of water in one or another fluid compartment. If, for example, elimination stops, even for a short time, there is a build-up of body water that overloads the system. *This can alter the personality.*

Excess body water causes tissue tension, swelling, and nervous irritability. When the Moon's gravitational pull upsets our fluid balance, the result makes us tense and liable to emotional outburst. A familiar example of the effect that build-up of body water has on nerves and behavior is the premenstrual-tension syndrome. Mood changes and increased irritability are well-known aspects of this syndrome, which can be used as a model of the biological high tide in the human body. During their premenstruum, women have increased frequency of visits to hospital emergency rooms, of admission to psychiatric hospitals, and of commission of crime.

Biological high tide sets the human machine on edge. It is important to note, however, that these tides do not *cause* strange behavior. They only make it more likely to happen. In persons already predisposed toward aberrant behavior or violence, the biological high tide can act as a stress trigger that releases irrational outbursts. Cosmic forces often act in a subtle manner. In certain situations they release energy of unexpected magnitude. We have learned that in some cases human beings may be susceptible to just such a trigger, in the form of the lunar-driven biological tide.

Our initial research findings were published in the July 1972 issue of the *American Journal of Psychiatry*. The findings were seized upon by the media and disseminated in the United States and abroad. Predictably, the public reaction was enthusiastic. After all, our studies confirmed the experience of people everywhere. I received a massive correspondence; personal lunar anecdotes were shared. The attitude of most scientists was open-minded and expectant. Many wanted to initiate studies of their own in efforts to replicate and extend our findings; articles confirming our findings appeared in the scientific literature. Several doctoral candidates chose our controversial subject for their thesis work.

There was, however, an attitude of skepticism on the part of "high science," and we waited for the inevitable "scientific" study refuting our findings. We had to wait only two years.

Alex D. Pokorny, M.D., and Joseph Jachimczyk, M.D., in an article titled "The Questionable Relationship Between Homicides and the Lunar Cycle," in the July 1974 issue of the *American Journal of Psychiatry*, used statistics developed in Houston, Texas. Because they were unable to demonstrate a statistically significant lunar periodicity in their data, the two concluded that "the effect of moon phases on homicide, suicide, and mental illness should be viewed as a myth."

Reviewing their research, I discovered that Drs. Pokorny and Jachimczyk used an inappropriate variable: *time of death*. They justified their use of time of death because 85 per cent of homicide victims in the Houston area were asserted to have died within one hour of the time of injury. Even if this were correct, it would still represent a 15 per cent error, consistently in the same direction in time (death invariably occurs somewhat later than injury). Any statistician would confirm that such an error plays havoc with the results of a survey designed to detect a subtle effect.

In order to demonstrate once and for all the fallacy of using time of death as a variable, we ran the Dade study again. This time we used time of death rather than time of injury as the measuring rod. We also reassessed the Cuyahoga data and added to our work a sample consisting of ten thousand homicide cases from New York City, which only listed time of death. In each of the three studies, there was no significant deviation from chance expectation. In other words, *violent injuries show a lunar periodicity*, but deaths are distributed randomly throughout the lunar cycle.

It was not surprising that Drs. Pokorny and Jachimczyk claimed to have found no relationship between homicides and the lunar cycle. More often than one might expect, scientists design studies to support their own private opinions rather than to document objectively what is going on. Many such studies have used inappropriate variables or inadequate data samples or the wrong time periods or lacked replication. Such reports only serve to continue controversy and to hamper progress toward resolution of the pressing scientific question. There is clear prejudice against work showing the Moon's influence on daily life.

Why the continuing prejudice? Science has always opposed the credulous acceptance of superstition. However, this attitude has been extended by some to a refusal to examine *any* "nonscientific" belief. An unconscious fear of the power of superstition over the mind may lie behind this refusal. Scientists by now should be able to see the difference between investigating a belief and credulous acceptance.

Lunar culture, with its poetic and mythic traditions, is not pleasing to the rationalist mind. However, we can no longer bury our head in the sand. The Moon will not go away. Its power will not suddenly cease to influence life.

There is another source of prejudice. Much of the skepticism concerning lunar influence on human behavior stems from the fact that Western science is time-locked into the solar cycle. This coupling means design of behavioral experiments is based on the regular monthly calendar and the twenty-four-hour clock we live by. (The Gregorian calendar is a solar calendar.) This sun-sided temporal orientation overlooks lunar influences simply by ignoring lunar timing.

Lunar time is out of phase with solar time. The lunar day is about fifty minutes longer than the solar day. This means that a lunar-timed event happening at, let's say, 8 A.M. today will occur at 8:50 A.M. tomorrow and at 9:40 A.M. the following day, and so on.

My work demonstrates a biological rhythm of human aggression; other research shows a biological rhythm in human sexuality. Both rhythms have been shown to resonate with the *lunar cycle*. Because aggression and sexuality are basic drives underlying human and animal behavior, we can conclude that many aspects of behavior reflect the net effect of *both* solar and lunar timing. When general science accepts this premise, an enormous amount of earlier research will be invalidated for the simple reason that a crucial variable—lunar timing—was overlooked. Science will eventually face these facts and take them into account in the design of future research.

It is well known that scientists have met with difficulty in replicating behavioral studies done in different places and at different times. I believe failure to take lunar time into account is partially responsible for the difficulty. The scientist who has many years invested in behavioral research can be resistant to accepting an idea that suggests he didn't have a handle on what he was studying.

Bearing this in mind, it was not surprising that Drs. Pokorny and Jachimczyk had concluded that "the effect of moon phases on homicide, suicide, and mental illness should be viewed as a myth." Their conclusion could hardly be justified by the results of a single study that used an inappropriate variable different from that of the study it purported to refute. *The attitude and the choice of words were those of scientific dogma resisting change—as if by reflex action.*

There is new scientific confirmation of the existence of a generalized lunar rhythm of violent deaths. Under a grant from the National Institute of Mental Health, Dr. Edward J. Malmstrom of Berkeley's Wright Institute found statistically significant lunar periodicity among homicides and suicides that had occurred in Alameda County, California, and in Denver County, Colorado, for the same fifteen-year period (1956 to 1970) as the Dade and Cuyahoga studies. He used time of injury as his measuring variable, adhered closely to my computer methodology, and adapted his statistical tests to the data. Filed with NIMH in December 1977, Dr. Malmstrom's work represents successful and significant replication, in two additional geographical locations, of my pioneer work.

3

The Moon
and Aggression

Demoniac frenzy, moping melancholy,
And moon-struck madness.
                    —*Paradise Lost*

Because the Moon affects the lives of everyone, it should be possible to demonstrate its influence on behavior other than the most drastic of violent acts. To expand the scope of our investigation beyond murder, we decided to collect data on the following violent events: suicides, aggravated assaults, fatal traffic accidents, psychiatric emergency-room visits. If we could find evidence of lunar rhythms in their occurrence as well, our concept of a lunar influence on human aggression would be strengthened. We also wished to test our assumption that lunar effects were due to the direct and indirect actions of gravity. For this purpose, we needed to study the effect on human behavior of lunar cycles other than the synodic, or moon-phase, cycle.

The Moon orbits the Earth every 29.5 days. During the course of orbit, it reflects more or less light of the Sun. When the Moon is between the Sun and the Earth, it reflects no sunlight toward the Earth and there is what is called the new-moon phase, or "dark of the moon." When the Moon is on the far side of the Earth, it reflects full sunlight to our planet and we have full moon.

The Earth is not the only point of reference for the cycles of the Moon. The Moon, in fact, has about four hundred cycles. In terms of the gravitational force exerted on the Earth, the only cycles of importance are the synodic cycle, the apogee-perigee cycle, the daily lunar-transit cycle, and the eclipse cycle.

Because solar and lunar eclipses occur no more than twice each in a year, it is not possible to collect ample data to test the eclipse cycle by itself. It is possible, though, to examine the apogee-perigee cycle and the daily lunar-transit cycle.

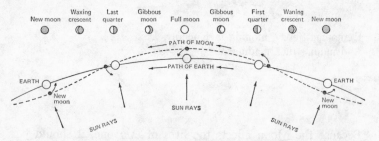

Paths of Earth and Moon during lunar month.

Apogee occurs when the Moon reaches its farthest distance from Earth and perigee occurs when the Moon makes its closest approach. (Apogee-perigee is a 27.5-day cycle.) Because the force of gravity varies with the inverse square of the distance between two attracting objects, the strength of the Moon's pull is increased at perigee. Times of lunar apogee and perigee are charted in the *American Ephemeris and Nautical Almanac*. We retrieved the exact times of apogee and perigee for the fifteen-year period encompassed by our study. Minor modifications of the computer program allowed us to test our homicide data in relation to the apogee-perigee cycle.

Times of high gravitational impact on the Earth occur when there is a coincidence of new or full moon with lunar perigee. During the fifteen-year study period, coincidence of lunar perigee with new or full moon occurred twenty-nine times. "Time-window" analysis of the frequency of murders during these twenty-nine periods yielded negative results. There was no significant deviation from chance expectation.

At first, this seemed to contradict my hypothesis. Had we relied on statistical correlations alone, this might have been a fatal blow to the biological tides theory, and we might have gone no further. Fortunately, we did not limit our work to the computer room.

During the five years of our research program, we were constantly attuned to local and national news media, and I personally spent many hours at the medical examiner's checking for increases of violent death. In late August 1970, shortly after the study began, there was an onslaught of shocking murders in the Miami area. Coincidentally, many brutal crimes, some of which were sex related, were reported from around the country. I consulted my ephemeris—the table that lists the dates of future astronomical events—and found that a major coincidence of cosmic cycles was underway. Full moon, lunar perigee, and an eclipse were combined so that the Sun, Moon, and Earth were aligned in a straight plane and the forces of gravity were greatly increased. The autumnal equinox occurred within the next two weeks, compounding the gravitational stress. During such a rare coincidence of cycles, dramatically high tides are recorded at ocean shores. I assumed biological tides would be affected. They were!

The murder rate increased significantly in Dade County at this time of gravitational stress. In September and October of 1970, the number of homicides was double the usual monthly average. Many of the murders had a strange, bizarre quality. One way to illustrate what happened is to give a few examples of the type of local crime that occurred during the period.

On September 23, 1970, Socrates "Shorty" Johns, a retired racing driver, was working in his son's tire store when three armed men entered and held him up. Although unarmed, Mr. Johns resisted. He was shot and wounded. He ran from behind the counter and chased the robbers out of the store and down the street, whereupon one of them turned and fired again, killing the man.

On the same day as the Johns murder, Gyorgy Virag was

leaving a restaurant with three companions when they were confronted by two men with a shotgun. One shouted, "This is a holdup." Mr. Virag yelled, "I'll get my gun," and ran toward his car. He was clubbed by one of the assailants, then shot in the head.

The next day, Debbie Oldham, twenty-three years old, was at her job as barmaid at the Casino Bar when two men came in and ordered sandwiches. When one of the men suddenly drew a pistol, Miss Oldham reacted hysterically, waving her arms, running about, and screaming. She was shot four times by the panicky gunman.

Three of the victims, when in a critical life-or-death situation, behaved in such a manner as to invite violence against themselves. It may be that gravitational stress precipitates self-destructive behavior as well as outward aggression. The two activities are certainly linked, as depth psychology has shown.

Faced with this obvious and dramatic increase in disturbed behavior at coincidence of *three* cosmic cycles, we were perplexed. Why had we failed to find an increase in murder at coincidence of *two* cycles—new or full moon and lunar perigee? The situation presented a frustrating logical contradiction. Could it be that more than two of the Moon's cycles had to coincide for there to be a significant increase in killings?

According to my theory, we should expect an increase in violent behavior whenever any increase in gravity above normal occurs. The correlation of murders in Dade County with new and full moon showed that *one* lunar cycle could influence the incidence of violence. How could we account for the subsequent contradictory findings?

The solution to the dilemma was found during one of my visits to the laboratory of Dr. Frank Brown in Evanston. I had met Dr. Brown at international scientific meetings and we became informal collaborators upon recognizing the parallel nature of our research. Dr. Brown studies plant and animal life for the effect of geophysical variables on their activity. I reviewed the recorded rhythmic activity of his "subjects" relative to moon phases. Graphs of the metabolic activity of bean seeds, potato sprouts, and hamsters showed a regular periodicity, with peaks around new and full moon. While reviewing a chart of hamster activity, I noted that occasionally one of the new- or full-moon peaks would be dramatically increased. On other occasions, one of the peaks would be totally reversed; activity was greatly *decreased*.

I asked Dr. Brown about the disparity, and he told me that *all* organisms showed similar periodic increases and decreases. He had concluded, after many years of observation, that organisms are either in a state of positive or negative receptivity relative to their natural environment. If one is in a state of positive receptivity, any significant environmental perturbation will cause an augmentation of peak activity. If one is in a state of negative receptivity, environmental perturbation will cause a phase reversal and the organism's usual peak of activity then assumes a downward trend. Positive or negative receptivity seemed to be a matter of chance. Roughly half the time an organism will be in one state and roughly half the time in the other.

I reasoned that we humans experience the same states of receptivity. There is no reason to think that we differ from plants and animals in our receptivity to environmental influences (though we certainly have more alternatives for dealing with them). This probability would explain why it was not possible in a single experiment to show an increase in the *number* of murders during certain times of extragravitational stress. It would also explain the exceptionally bizarre nature of crimes that were committed at such times. Unbalanced individuals who are in a state of positive receptivity might be inclined at these times to show greater excess in the nature of their violent outbursts. Simultaneously, of course, about half the population will be in a state of negative receptivity; these people would be exceptionally quiet and possibly withdrawn.

Many people feel they are most calm and relaxed at new or full moon. Some consider these periods to be optimal for passive pursuits such as meditation. Psychic phenomena are said to be augmented or enhanced at full and new moon. The receptivity factor provides a clear example of how reliance on statis-

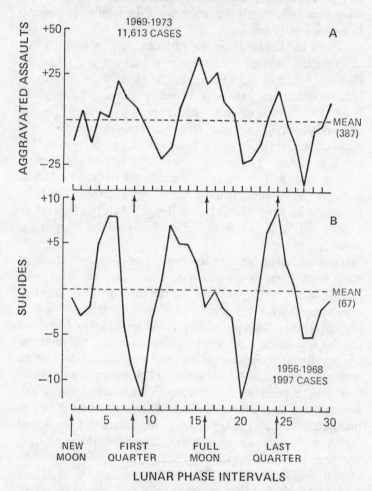

AGGRAVATED ASSAULTS AND SUICIDES PLOTTED IN
RELATION TO THE LUNAR SYNODIC CYCLE*
DADE COUNTY, FLORIDA

* The data was smoothed using a 3-point moving average. The curves are plotted about the mean. Note the similarity between the two periodicities relative to the lunar synodic cycle.

tics can be misleading. If at any given time half the population is liable to become agitated and the other half quiescent, the average number of violent incidents will not be greatly changed. However, those incidents that do occur at such times tend to have a dramatic quality about them. This concept helps to explain why, in investigating lunar influence, we are dealing (from a statistical standpoint) with small effects.

Dr. Brown long ago gave up reliance on statistics for documentation. He has far more confidence in detailed long-term observations. Other researchers in lunar effects agree with Dr. Brown. Statistics can only provide a cross-sectional picture for a single point in time. This picture is static and in no way reflects the dynamic interaction between organism and environment. Statistical methods that delineate periodicity are useful but limited in scope.

We examined our data samples relative to the daily lunar-transit cycle. The lunar-transit cycle governs the day's two low and two high tides. The cycle results from the rotation of the Earth on its axis, relative to the Moon and to the Sun. Upper and lower lunar transit marks the two times every day that the Moon exerts maximum attraction. Investigation revealed no correlation between murder and the lunar-transit cycle.

It became clear to me that the moon-phase cycle is the key lunar cycle in the timing of plant, animal, and human activities. The other lunar cycles are of little consequence in and of themselves. However, at certain points when one or more of them interacts with the synodic cycle, they contribute to lunar effects. (Meteorologists report similar findings. The synodic cycle is the most important lunar cycle in the timing of lunar effects on weather.)

As noted earlier, we decided to extend the study by examining aggressive behavior other than homicide. We obtained Dade County records of the incidence of aggravated assaults, suicides, fatal traffic accidents, and psychiatric emergency-room visits. The aggressive nature of the emergency-room visits is debatable; for our sample it was established that 80 per cent of the "visitors" behaved in a way that was dangerous to themselves or others. One-fourth had attempted suicide.

For our various behaviors under study, we collected data covering periods from five years to fifteen years. With one exception they were screened to eliminate cases with imprecise timing. The exception was suicide. Usually there is no witness to a suicide. The person is discovered some time after death. For suicides, time of injury is estimated through police investigation and autopsy findings. We had to discard many cases of suicide because the estimates were obviously unreliable. We included *properly* documented suicides because recent studies of suicide attempts showed a lunar influence.

When we tested each type of behavior relative to the moon-phase cycle, a significant correlation was found in every case with the exception of suicide. In spite of this exception, it appears likely that suicide has some intrinsic relation to moon phases because other destructive behavior shows a clear-cut lunar periodicity. The inaccurate suicide data are probably responsible for the negative results. A better way to study the frequency of suicide must be devised.

Aggravated assaults, which are regarded as essentially the same crime as homicide, except that the victim survives, peaked around full moon. There was a secondary peak shortly before new moon. The peaks were consistent with the homicide findings. Violence directed against others showed a confirmed predilection for full moon, with somewhat lesser clustering around new moon. Fatal traffic accidents peaked between first quarter and full moon and again at last quarter. Psychiatric emergency-room visits peaked around first quarter and last quarter, with a significant decrease at new moon and full moon.

In an earlier survey of psychiatric-hospital admissions, Dr. Roger Osborn noted that, "If the moon phase does influence psychiatric hospital admissions, any moon-influenced disturbance might not be recognized as serious enough to warrant hospitalization until it had existed for a few days. If so, the higher admission rate would trail the actual moon phase." If psychi-

---

* The data was smoothed using a 3-point moving average. The curves are plotted about the mean. Note the similarity between the two periodicities relative to the lunar synodic cycle.

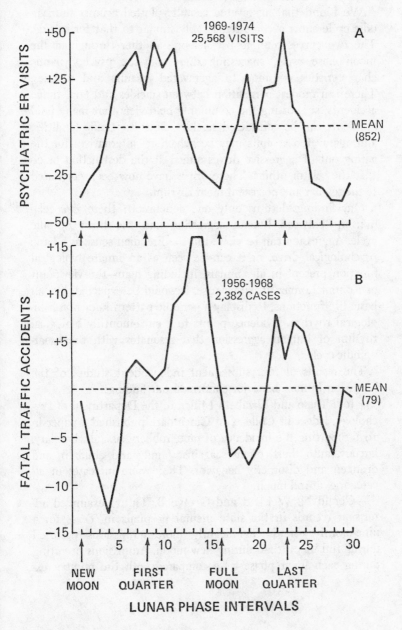

PSYCHIATRIC ER VISITS AND FATAL TRAFFIC
ACCIDENTS PLOTTED IN RELATION
TO THE LUNAR SYNODIC CYCLE*
DADE COUNTY, FLORIDA

A

1969-1974
25,568 VISITS

PSYCHIATRIC ER VISITS

+50
+25
MEAN
(852)
−25
−50

B

1956-1968
2,382 CASES

FATAL TRAFFIC ACCIDENTS

+15
+10
+5
MEAN
(79)
−5
−10
−15

5   10   15   20   25   30

NEW
MOON

FIRST
QUARTER

FULL
MOON

LAST
QUARTER

LUNAR PHASE INTERVALS

atric emergency-room visits were subject to the same lag, it might account for the delayed peaks in this case.

We found that aggravated assaults plotted against the synodic cycle show a curve remarkably similar to that for suicide. The two curves tend to parallel one another throughout the moon-phase cycle, suggesting there may be psychodynamic characteristics common to aggravated assaults and suicides. There is a modest correlation between suicides and fatal traffic accidents, supporting the commonly held view that many fatal accidents can be considered suicides. Suicides, traffic fatalities, and aggravated assaults may be behavioral alternatives for the acting out of aggressive drives aimed at the destruction of either the self or others. These drives have now been correlated to the regular and persistent lunar rhythm.

Our findings left us only one conclusion: there is a relationship between human aggression and the lunar synodic cycle. Aggression can be viewed in the Freudian sense as a basic psychological drive, or it can be seen as an innate biological function present in all animals, including man—the viewpoint of Konrad Lorenz. In either case, it would be expected, like all basic life functions, to display a periodic pattern known as a biological rhythm. Evidence points to a circa-monthly biological rhythm of human aggression that resonates with the lunar-synodic cycle.

The results of four subsequent independent studies of the Moon on human behavior confirmed our findings.

—Jodi Tasso and Elizabeth Miller, of the Department of Psychology, Edgecliff College, in Cincinnati, published a paper in 1976 reporting the incidence of rape, robbery, assault, burglary, larceny, auto theft, offenses against family and children, and drunken and disorderly behavior. There were increases in *all* categories at full moon.

—Gerald N. Weiskott and George B. Tipton examined admission records in the state mental hospitals in Texas for a nine-month period. Significantly more admissions occurred during full moon than during new moon. Admissions occurring during each lunar phase were compared with the number expected by chance. The larger percentages occurred at full moon and the last quarter.

—Sheldon Blackman and Don Catalina studied admissions to the emergency room of a community mental-health center in Staten Island, New York, for a one-year period. There were significantly more admissions on full-moon days than at other times during the month. Comparing their data with certain weather variables, Messrs. Blackman and Catalina could find no relationship between visibility of the moon and the higher rate of admissions at full moon.

—Klaus-Peter Ossenkopp and Margitta Ossenkopp found a lunar periodicity in self-inflicted injuries among a population of female psychiatric outpatients in Canada.

Armed with our expanded research findings and several years' experience, we were in a position to risk predicting times of behavioral upheaval. In the late summer of 1973, I came across an article in the science section of *Time* magazine reporting that a major coincidence of cosmic cycles, resulting in unusually high tides, was due to occur in January and February of 1974. The Earth, the Moon, and the Sun would be in a straight line, a position called syzygy, with the Moon at a perigee unusually close to Earth. In addition, perihelion would occur, with the Earth making its closest approach to the Sun. In 1962 unusually high tides caused forty deaths and $500 million in property damage along the mid-Atlantic seacoast alone. The ocean was whipped by gusts of up to seventy knots and rose 9.5 feet above the average low-water mark. (We have already indicated the dramatic behavioral consequences of a similar planetary alignment that occurred in September 1970.)

I alerted my listening posts in the Miami police department, the local news media, the psychiatric emergency room in Jackson Memorial Hospital, and the Dade County Medical Examiner's office. We would attempt to document any increase in the types of behavior previously investigated. In a letter to the medical examiner, Dr. Joseph Davis, I predicted a general disturbance in human behavior during the period of the impending cosmic coincidence. I predicted an increase in visits to emergency rooms, a surge in psychiatric hospital admissions, an increase in accidents of all kinds, and an increase in the number of homicides. I warned Dr. Davis to be on the lookout for murders of "a qualitatively bizarre and brutal nature, including victim-provoked murders." I also suggested the practical import of the experiment: ". . . if, in fact, we are able to predict even rare periods of general behavioral upheavals, we may in the future have something to offer in terms of preventive approaches."

Having risked committing myself to a prediction on paper, I eagerly awaited the major coincidence of cosmic cycles.

Sure enough, all hell broke loose, especially during the first two weeks of January, with the Moon rocking only 217,000 miles from the Earth. The oceans were unusually restless. On January 8, coastal California was ravaged by an incredibly destructive perigean spring tide. The direction of the wind amplified the effect.

Back in Miami, SHARP RISE IN MURDER RATE SHOCKS CITY POLICE was the headline in the Miami *News* on January 21. The murder toll for the first three weeks of the new year was three times higher than for all of January 1973. It caused deep concern among police, particularly because of the brutality of most of the crimes and, in some, the lack of obvious motive. What follows is a partial listing of the bizarre and brutal events that occurred during this period of exceptional cosmic stress. During the first two weeks of January alone, there were nine brutal murders in and near Miami:

—In Orlando, two police officers were slain in separate incidents.

—Samuel Carson was talking to friends outside his apartment house when suddenly—and apparently without reason—two shots rang out, one of which fatally wounded him.

—On the same night as the Carson murder, a seventy-year-old woman was savagely beaten and raped in a cemetery. She died the next day.

—A forty-five-year-old man stabbed his twenty-five-year-old son to death after an argument.

—A fifty-seven-year-old drifter was beaten with a board and his throat was slashed in broad daylight, and apparently without motive, in a littered lot near downtown Miami. ("It's unusual for a wino to be killed like that," said a police homicide sergeant.)

—A thirty-two-year-old medical resident at Jackson Memorial Hospital was bound, beaten, and strangled in his apartment.

—A man shot and killed his wife, then turned the gun on himself in the back seat of a moving car. (In the front seat were the couple's two children and the man's brother-in-law, who was driving.)

During this time of cosmic coincidence, episodes of excessive and bizarre violence were of course not confined to the Miami area. Among many concurrent brutal murders around the world were:

—In a New York suburb three teen-agers kidnaped a fourteen-year-old boy and demanded ransom. Though $15,000 was paid, they strangled their captive. (The boy's body was found tied to a tree in a snow-covered wooded area.)

—In Taos, New Mexico, an unemployed farm worker shot each of his four children in the head. Three died.

—In San Joaquin, Brazil, a woman used a machete to open the chest of her ten-month-old son in order to free a demon. (The murder was committed in the presence of the fourteen other members of the family.)

—In Wichita, Kansas, four members of a family were murdered in their home. Each victim was severely beaten, then killed. (One had a plastic bag tied over the head. Two were strangled. The fourth was hanged from plumbing pipes in the basement.)

Not only humans were affected by the excessive tidal stress. I received reports of a one-ton African water buffalo, generally thought to be a mild and compliant beast, going berserk, goring and trampling a game warden to death.

Fortunately, a major coincidence of cosmic cycles is rare. Fergus J. Wood, a researcher with National Ocean Survey, predicts that the next such coincidences will occur in December 1990 and in January 1992. We have a long time to prepare.

Deaths due to all causes significantly increased in Dade County during the three-month period from January through

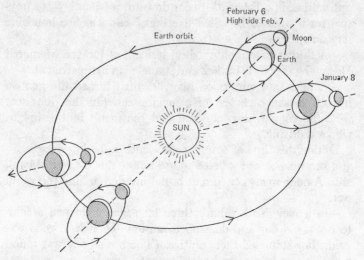

Alignment of Sun, Earth, and Moon resulted in destructive perigean spring tides in 1974. Similar alignment will occur in 1990 and in 1992.

March 1974. Psychiatric emergency-room visits took a dramatic leap, increasing in January alone by 35 per cent over the previous month. For the full three-month period observed, psychiatric emergency-room visits were almost 40 per cent higher than during the same period of the previous year. Admissions to the psychiatric institute at Jackson Memorial Hospital were at near capacity during the three months; the average daily census there was significantly higher than for the same period of the two previous years.

We were not able to predict precisely which types of violence would increase or how much the increase would be. We were able, though, to predict in a general sense a massive disturbance on the basis of a coincidence of cosmic cycles.

The outburst of violence in the first three weeks of January 1974 was accurately predicted. A tailing-off of the violence made statistical verification difficult, if not impossible. From a purely statistical standpoint, only the emergency-room visits were significantly increased. However, Dr. Davis, city police officials, and members of the news media with whom I shared my prediction were convinced of its accuracy. An early-warning system of preventive psychiatry and criminology for the future was now a real possibility.

4

The Moon and Natural Cycles

And as for the revolution of these heavenly bodies, there may very well be other principles which lie behind them. Nature's aim, then, is to measure the generations and endings of things by the measures of these bodies, but she cannot bring this about exactly on account of the indeterminateness of matter and the existence of a plurality of principles which impede the natural processes of generation and dissolution and are so often the causes of things occurring contrary to nature.

—Aristotle

The moon is nothing
But a circumambulatory aphrodisiac
Divinely subsidized to provoke the world
Into a rising birth-rate.
                —*The Lady's Not for Burning*,
                Christopher Fry

Since ancient times, it has been apparent that life itself is a rhythmic process. As Aristotle noted, the rhythms of life are not always exact nor are they always obvious.

Some of the familiar rhythms that have been examined in more recent times are the cycles of sleep-waking, feeding, phys-

ical activity, and sex and reproduction. These cycles are predictable from individual to individual within a species, although
the precise timing of the cycle will vary. We all know people
who are definitely "night people" or "day people" and that this
timing is an expression of their physical-activity cycle. Eating
cycles vary widely. Some people put away a Brobdingnagian
breakfast, while others are nauseated by the smell of food on
arising in the morning.

In times past, these cycles were remarked upon and linked to
the movement of the heavenly bodies, but modern science was
slow to understand and accept them. Reports documenting
biorhythms became popular in the 1880s and 1890s. In 1898 a
paper documenting many biological rhythms was published by
the Danish Nobel Laureate Svante Arrhenius. Biological cycles
were given the name *circadian rhythms* by Dr. Franz Halberg
and his associates at the University of Minnesota some thirty
years ago. Circadian means "around a day"; it is used to refer
to the daily repetition of many cycles. Other biological cycles,
such as the menstrual cycle, have long periods of repetition.
Others, like the cycles that run within the normal period of
sleep, are of short duration. Cycles have been demonstrated in
a wide variety of human functions—temperature, heartbeat,
sensitivity to pain, color sensitivity, and so on. They persist in
people who have descended into a deep cave and lived there for
months. Man is a symphony of rhythms and cycles.

Biological rhythms can be disturbed by influences like
trauma, disease, heat, cold, and the weather. The now familiar
jet lag is an example of the way normal activity rhythm can be
upset by long-distance travel.

Recently, it was discovered that any one of four substances—
heavy water, dalinomycin (an antibiotic), alcohol, and lithium
—can shift biological rhythms. Rhythms also can be shifted or
entrained by the use of the artificial stimulus of light when the
subject is isolated from sunlight. Though shifted, rhythms will
continue their regular functioning.

Even under laboratory conditions, and isolated from the
light of the Sun and the Moon, there are solar-day rhythms (24
hours), lunar-day rhythms (24.8 hours), lunar-synodic monthly

rhythms (29.5 days), and yearly rhythms. Because these rhythms persist under what were thought to be constant and isolated situations, they were presumed to be under the control of some sort of internal clock. The majority of researchers in the field of biological rhythms subscribes to the theory of the biological clock. This self-timing clock is thought to be present, somehow, within each and every cell of every organism. The clock is said to set the individual's rhythms and to maintain them independent of outside influences. At present, however, nobody has had any success in locating the mechanism of the mysterious and elusive biological clock.

A second theory of biological rhythms was espoused by my colleague Dr. Frank Brown, who is considered the chief proponent of the *extrinsic* timing theory of biological rhythms. The rhythms, he believes, pick up their timing from external cues, which are provided by cosmic cycles. Cosmic cycles trigger regular fluctuations in our environment. Human ecology resonates with these rhythms; the rising and setting of the Sun is an obvious example of this resonating pattern. The ebb and flow of daily tides is another; the tides resonate in a double period slightly longer than the solar day.

Light is the most obvious of timing cues. Dr. Brown has demonstrated that animals will delay or advance their periods of activity according to the gradual seasonal changes in the length of the day. In the laboratory, lighting conditions can be manipulated easily to demonstrate the operation of this cuing mechanism.

For animals living on the seashore, the rhythm of the tides provides key cues. The tides, for example, determine when it is best to be active and to feed. Fiddler crabs scuttle about and feed according to the tides, but their skin darkens and lightens in a pattern governed by the solar day. Two rhythms, lunar and solar, are at work in the same organism; a persistent semimonthly rhythm results from the interactions of the two circadian rhythms. Dr. Brown learned that the fiddler crab's rhythmic response does not depend on a complex clock mechanism but on natural cycles. The lunar response is present in land animals as a consequence of their evolution from marine to land-dwelling creatures. At one time, all earthly organisms were marine and influenced by the tides.

It is easy to see how light from the Sun can act as a timing cue, but how can the gravity of the Moon act as a cue? Shellfish in tidal waters could be getting their cues from the flow of the tides, or they could be getting them by some other means. To investigate tidal timing cues, Dr. Brown had oysters shipped by air from the Connecticut shore to his laboratory in Evanston.

The oysters were known to open their shells at high tide. In the laboratory, conditions were controlled as carefully as possible. No external cues were allowed to penetrate to the oysters in their pans of sea water. For the first week, the oysters opened their shells at the times of high tides at their ancestral beds in Connecticut. They were continuing their accustomed rhythm. After two weeks had passed, however, their timing changed. Now they were opening their shells at the time when the Moon was at zenith over their new home in Illinois. If Evanston were a coastal city, it would have been the time of high tides. The oysters must have been taking their cues directly from the gravitational pull of the Moon!

But how could they sense the pull of the Moon, shut off, as they were, from the "outside" world? The indication is that some lunar influence causes changes in the electromagnetic fields that surround *every* living organism. The oysters sensed the change and adapted to it. Researchers theorize that part of the nervous system receives field-change signals.

However they sense the pull, myriad sea creatures are affected by the rhythms of the Moon. Exact lunar rhythms have been demonstrated in the reproductive cycles of fish. The California grunion breeds only on nights immediately following full or new moon. At the waning of the moon, European eels start their migration to spawning grounds in the Sargasso Sea. The penultimate-hour crab is most active at high tide. The fiddler crab's activity occurs at low tide.

Some sea creatures show highly sophisticated lunar sensitivity. The palolo worm of Samoa and Fiji, a creature that lives almost all its life inside coral reefs, emerges to mate only on the night before the last quarter of the moon in October and November. Bursting with egg and sperm cases, the long tails of these worms break off and swarm to the surface, where the islanders scoop them up by the basketful. For the islanders, these are feast days, and the coincidence of the worm harvest with feast days is possible because lunar calendar timing is *precise*. Timing does not depend on maximum tidal pull, which would be stronger at full moon, but on a particular sequence in the lunar cycle.

Fishermen's traditional knowledge of the lunar timing of sea creatures has been modernized. A man named, appropriately enough, John S. Haddock produced a Fish Biting Hours Computer Calendar. According to this calendar, there are four daily periods of lunar feeding stimulus. These stimulus periods are never at the same daily clock time because the lunar day is somewhat longer than the solar day. The Haddock calendar converts lunar time to solar time and provides for geographical correction. It enables the fisherman to tell by the clock when his chances are best. Mr. Haddock claims most record fish catches are made during a lunar feeding-stimulus period. It is well to note that solar time alone would offer no clue at all to the useful feeding rhythm of fish.

The *extrinsic*, or cosmic cycle, timing theory of biological rhythms developed by Dr. Brown expresses a minority view. But many biorhythm researchers and scientists in other fields are beginning to accept and employ his ideas in the design of their own work. In my opinion, Brown's is the simpler and more elegant of the two theories of biological rhythms. It does not, like the *intrinsic*, or biological clock, theory, depend on a device for whose existence there is no evidence. The extrinsic timing theory depends on forces and systems whose existence and operations are well known: cosmic cycles, senses, nervous systems, and so forth. My own research relative to moon phases and human behavior tends to support the extrinsic timing theory of Dr. Brown.

Among the natural cycles whose timing is associated with the Moon, the reproductive cycle is prominent. So often we associate a moonlit night with love and romance. Does the romantic cliché represent a biological principle, another connection to the mysteries of the Moon? Is the full moon more than scenic inspiration for lovers? It is—at least to sea urchins. Their reproductive cycle follows the lunar cycle exactly.

All over the world, people believe there is a connection between the Moon and the human reproductive cycle. Is it only superstition? In India, for example, it is believed the phase of the moon under which a child is conceived will determine the sex of the child. It is also believed there are more childbirths on full- and new-moon days. The Navajo believe there are more births at full moon because of the Moon's pull on the embryonic fluid—a kind of biological tide.

Scientific research is beginning to catch up with folk beliefs. In 1961, Dr. Hilmar Heckert, a Berlin physician, demonstrated lunar rhythms for humans in several categories including births and deaths. In a study published in the *Journal of Genetic Psychology*, in 1966, Robert McDonald reported there were significantly more births during full and new moon in his data sample.

Another area for scientific research has been the possible connection between the menstrual cycle and the lunar month. These two cycles always have been believed to have about the same time span. Many scientists, including Dr. Heckert, claim to show a relationship between the timing of the menstrual cycle and the lunar month. Arrhenius reported on this connection in his seminal paper on biological rhythms in 1898. Similar studies were published by Romer (in 1907), Bramson (1929), and H. Guthmann and D. Oswald (1936); they showed a significant number of menstrual cycles started at full or new moon.

In 1959, Walter and Abraham Menaker published the results of a study of the timing of human reproduction in relation to the lunar cycle. They used a massive data base. Examining existing studies on the duration of the menstrual cycle, they found that, on the average, the human menstrual cycle is not approximately the same length as the lunar month—*it is exactly the same length*, twenty-nine and one-half days.

To examine the human gestation period, they used records of 250,000 births. The length of gestation is *precisely* nine lunar months—265.8 days or, to the nearest whole number, 266 days. They concluded that the human reproductive system follows lunar time rather than sidereal time. More births, it was documented, occurred at full moon than at other phases—another statistical proof of a popular belief once considered superstition. It still is not known whether more births occur at full moon because of lunar timing of conception or of the gravitational effect of the full moon on the birth process—a hastening of labor, as the Navajo believe.

Speculating on what the connection between the Moon and the body might be, the Menakers suggested that man may be subject to biological tides: "As our bodies are about two-thirds 'sea' and one-third 'land,' we must sustain 'tidal' effects." This speculation, in a field of investigation totally different than my principal area of violent behavior, is a significant reinforcement of the biological tides theory.

In fact, the Menakers' conclusions led them to believe, as Dr. Brown and I do, that the reproductive process is timed by the extrinsic astronomical clock rather than by a mysterious intrinsic biological clock.

Given the likelihood of a lunar-timing effect on the human reproductive cycle, one might wonder why it is that all women don't menstruate at the same time, especially (as is well known) as groups of women living together tend to have their periods at the same time. In presuming there is a lunar component of human menstruation, we must recognize there is a solar-time component as well. If all women were born at the same time and on the same day, it would be perfectly reasonable to expect they would all have their periods at the same time as well. This isn't the case, of course.

One can speculate how the lunar-timing component comes into play by tracking an entire reproductive cycle. While an embryo is developing within the womb, the biological rhythms of the fetus are in equilibrium with those of the mother. At the moment of birth, the newborn comes into contact with the natural geophysical environment in which it will spend the rest of its life. It is reasonable to assume that the rhythms of the newborn at this point are no longer dependent on those of its mother. They now become entrained by forces in the natural environment.

The independent biological rhythms of each human being are established at the time of birth, when the organism comes into contact with its environment and begins to be influenced by forces in that environment. Many of these forces are known to serve as cues for biological rhythms. These cues are imprinted at the moment of birth. Solar and lunar rhythms will be identifiable at any point in the life of the individual. The timing of the rhythms will differ from one individual to another because we are all born at different times and different places.

Because of the complex interplay between hormonal balance and fluid and electrolyte balance, women at certain times during the menstrual cycle may be more susceptible to the triggering effects of biological tides. During premenstruum, many women experience bloating, tension, and irritability. Transient fluid buildups and electrolyte imbalance occur. The added physiological stress of a biological high tide occurring during premenstruum could result in drastic physical and/or behavioral consequences. It is well documented that women are more susceptible to physical illness and erratic behavior in the few days prior to menses, and that hospital admissions, psychiatric illness, and violent crime are far more frequent then. Male hormones are less involved with fluid and electrolyte balance, and there seems to be no male counterpart of the premenstrual-tension syndrome.

If my theory is true—the environmental forces that derive ultimately from cosmic cycles are encountered at birth and thereupon set the individual's rhythms—we should expect to find evidence of the effect. The evidence would be found in the correlation of the cosmic configuration at the birth of an individual with the later biorhythmic behavior of the same person.

Just this sort of correlation has been made by Dr. Eugen Jonas, of Czechoslovakia. He believes the position of the Sun and the Moon when a girl-child is born determines the biorhythms of her reproductive system, including ovulation. With precise knowledge of her rhythmic ovulation pattern, a woman can practice birth control on the days she is likely to conceive. Jonas claims his astrobiologic system of birth control is 98 per cent effective. According to the Jonas system, which is quite different from the standard rhythm method in which intercourse is avoided at mid-cycle (the supposed time of ovulation), ovulation takes place twice each month, once at the sun-moon configuration at the time of the woman's birth and again at the configuration that is 180-degrees opposite. The times of ovulation are therefore different for each successive month. Ovulation, in effect, travels in a predictable progression across the menstrual cycle, based on the regular progression of the Sun and Moon.

Edmond M. Dewan, working at Data Sciences Laboratory, of the Air Force Cambridge Research Laboratories, L. G. Hanscom Field, proposes the use of extrinsic timing of biological clocks as a method of birth control. He found that artificial light could be used to regulate the menstrual cycles of women with irregular cycles. His results indicated the menstrual cycle may be entrained and regularized by the use of light, making the rhythm method of birth control foolproof, or almost. The light in his experiments, Mr. Dewan believes, affects ovulation through the hypothalamus, pituitary, and pineal glands. Further research is underway.

> The wavering planet most unstable,
> Goddess of the waters flowing,
> That bears a sway in each thing growing
> And makes my lady variable.
> Oft I seek to undermine her,
> Yet I know not where to find her.
> —Anonymous, sixteenth century

5

Cosmic Continuity

All life forms are cosmic resonators. We are an integral part of the living Universe. We are constantly influenced by the ebb and the flow of cosmic activity—moon phases, sunspots, solar radiation, cosmic rays, planetary movements, to list five of its forms. The fluctuations of these forces are subtle, yet regular and pervasive. They impinge on the Earth's atmosphere and cause perturbations in the environment at biosphere level. Changes are perceived by humans, animals, and plants, and are converted into biological rhythms detectable in inner electromagnetic activity.

The harmonics of this celestial dance are played in various modes. The Yale-based research team of Harold Saxton Burr and F. S. C. Northrup pioneered the study of "fields of life." Dr. Burr was a professor of anatomy for forty years* and Dr. Northrup, a professor of philosophy and law. They devised a method for measuring changes in the electrical potential of an organism. A sophisticated voltmeter was developed in order to measure electrical fields of organisms without drawing current from them. Over the years, electrical potential in plants, animals, and human beings was recorded.

Drs. Burr and Northrup referred to the fields as electrodynamic fields, or fields of life (L-fields). Hourly variations in the fields were noted and compared with variations in the fields of the Earth and air. There was a distinct parallel be-

* Shortly before his death in 1972, he published his *Blueprint for Immortality*.

tween life fields and the fields of the Earth and the air, indicating life fields varied in response to changes in the geophysical environment. The environmental changes paralleled the regular changes of the solar and lunar cycles. For example, analysis of field changes in trees near their laboratory in New Haven yielded reflections of solar and lunar cycles.

The professors' work revealed that the life field of a system was a sensitive indicator of biological activity. In some applications it could be used to predict biological activity, indicating to Burr and Northrup that the life field of a system regulates an organism's structure and function. This is some of their supporting evidence: field determinations in seeds predict growth patterns, size, and vitality of plants; measurements of unfertilized frog eggs reveal the head-tail axis of the future nervous system; ovulation, wound-healing, peripheral nerve injury, and even the growth of cancers have electrical-field correlates.

Using the Burr-Northrup principles, Dr. Louis Langman, a gynecologist, was able to detect electrical changes in women who subsequently developed gynecological cancers. Other gynecologists were able to show that electrical changes in the ovaries *preceded* ovulation and that ovulation, in fact, occurred in many women throughout the menstrual cycle rather than only at mid-cycle. (By his "astrobiologic" method, Dr. Eugen Jonas, the Czech, also detected a similar capricious timing of ovulation.)

Dr. Leonard J. Ravitz, a psychiatrist, applied the Burr-Northrup technique to electrical state changes in mentally ill patients in psychiatric institutes at Yale University, Duke University, and the University of Pennsylvania. He found mental illness was accompanied by a change in the electrical-field state of the patient. Often, a return to a normal-field state heralded a return to mental health. Dr. Ravitz investigated hypnotic states and found they produce alterations in the electromagnetic fields of the subjects—establishing a link between states of consciousness and electromagnetic fields.

In general, Dr. Ravitz found that disturbances in behavior are most often linked to the *intensity* of the subject's field. He discovered "the most dramatic voltage-rises in certain cases had preceded those rare occasions when torpid schizophrenic behavior shifted into animated spontaneous functioning." Physical problems, on the other hand, are usually related to the *polarity* of the life field (plus or minus).

Dr. Ravitz made still another important discovery: electric fields of humans vary in a cycle that either peaks or drops at new and full moon. The lunar-cycle times the upward and downward extremes, but it cannot be used to predict direction. Changes in electric-field potential are important indicators of mood. High readings indicate hostility and apathy. Low readings are associated with good-natured moods.

The discovery of wide mood swings occurring at new and full moon gives support to my finding that aggressive activity peaks at these phases. Volatility of temperament, connected with the personal electrical field, means that people are more likely to fly off the handle at such times. The tendency of field measurements to go up *or* down dramatically at new and full moon lends support to Frank Brown's idea of the positive or negative receptivity of an organism.

In making this connection, many questions spring to mind. What, for instance, is the relationship between lunar-timed changes in body fields and biological high tides in the fluid compartments of the body? Is it through the nervous system that fluid balance is influenced, or vice versa? We certainly know the relationship can work changes in both directions. We also know the changes affect personality and behavior. Why does a life field swing up or down dramatically? Why might one at any given full moon be in a state of positive or negative receptivity?

Experiments performed by Dr. Brown shed light on the mystery of life's receptivity to the electromagnetic environment. There is a continuum of fields between the individual and his surroundings. Electromagnetic fields have no distinct borders; they are always shifting. To demonstrate the reciprocal relationship of two neighboring organisms under changing electromagnetic influence, Dr. Brown used dried bean seeds. Why dried bean seeds? Simply because they are simple. When the seeds are immersed in a vessel of water, the uptake of water is their only metabolic function. They show a periodic pattern in their uptake of water, with peaks at new and full moon.

The degree of water-uptake by a bean seed can be changed by altering the course and direction of the electromagnetic field in which the bean seed is rotated. This change can be accomplished by the expedient of rotating a bar magnet below the table on which the bean seed is immersed. Change can also be brought about by placing the vessel containing the bean seed on a revolving platform and rotating the seed in a circular manner around the points of the compass. In effect, this places the seed in differing relationships to the magnetic field of the Earth.

When two bean seeds are placed in the same vessel of water, a remarkable relationship develops. One bean seed takes up the water while the other will not! By rotating the two seeds in a different magnetic field pattern, the "dry" bean seed can be made to take up the water while the "wet" seed will not. This experiment suggests that environmental changes not apparent to our senses can affect a reciprocal relationship between two members of the same species. It is implicit also that subtle electrical messages are transmitted between the two seeds. There is subliminal communication—one member of the pair agreeing to undertake the active metabolic role while its mate assumes a passive role.

This most elegant series of experiments sheds light on the basic nature of a reciprocal relationship between two members of the same species. The mysterious attraction (or repulsion) between two relative strangers has heretofore been labeled "vibes" or "chemistry." It appears that "physics" is perhaps more appropriate and "personal magnetism" may be getting even closer to the truth of such interpersonal phenomena.

Parapsychologists investigating extrasensory communication should be attuned to these findings of Dr. Frank Brown. Perhaps the bean seed relationship constitutes a prototype ESP. The evidence suggests the electrical systems of living things directly perceive electromagnetic fields around them. As Dr. Brown points out, this is inevitable because the two are continuous: "No clear boundary exists between the organism's metabolically maintained electromagnetic fields and those of its geophysical environment." Other researchers, such as Dr. Robert Becker of the Veterans Administration Hospital in Syracuse, New York, have also detected evidence that the neural electronic system acts as a receiver of electromagnetic information.

What is of immediate concern to my work is that organisms are probably capable of perceiving changes in the Earth's electromagnetic field brought about by the movement of the Moon in relation to the Earth and the Sun. This, in itself, is a sort of extrasensory perception.

Dr. Brown has warned that the uncertainty principle postulated by the physicist Werner Heisenberg will be operating more and more in biological experiments. Experimental measurements, proceeding by increasingly sophisticated techniques, will affect the biological processes being measured. We may be changing the very life we are examining. "Biological processes will reflect in their measured values the methods and conditions under which the measurements are made, and the differences may be substantial," says Dr. Brown.

The idea of the individual's continuity with the rest of the Universe is not a new one. Throughout history, saints and mystics have presented a variety of metaphors for the individual's feeling of being "one with the Universe," the "oceanic feeling." The knowledge that our internally generated electromagnetic fields are actually continuous with those of the environment provides a physical basis, however subtle, for this "Eternal Truth."

We should bear in mind that in the realm of consciousness a minuscule input can have gargantuan results. Experiments with electrical stimulation of the brain (ESB) have demonstrated that small changes in electrical potential in the brain are enough to reverse completely the behavior of an animal or a human. In the case of the electromagnetic field that surrounds us like a seamless web, the medium is overlooked while the message can be read in our moods, which may include violent outbursts. Continuity with environment is not always a happy affair.

Stress is one connection with our environment that we would like to avoid but are invariably forced to deal with. It has been well established that stress, both mental and physical, can make us more susceptible to subtle influences and forces. I have emphasized the point that it is those people who are predisposed toward violent aggressive behavior who are most likely to be affected by the Moon (such people generally have a lowered stress tolerance). Social, psychological, and physical stresses tend to set such people on hair-trigger. We should expect stress to be a significant predisposing factor in the lunar influence on human behavior. And so it is.

The stress connection was explored by biologist Harry Rounds of Wichita State University. He became interested in lunar-timed behavior when, in investigating blood factors in cockroaches, he found changes in their blood showed a pronounced relationship to the phases of the Moon! Intrigued, he resolved to look into the matter. He tested the blood of roaches, mice, and men in order to detect chemicals in the blood that cause the heartbeat to accelerate. Knowing that stress is a definite factor in the heartbeat, the biologist divided his experimental subjects into two categories—stressed and unstressed. Acceleratory factors in the blood of stressed animals rose sharply shortly *after* new and full moon! By way of contrast, cardioacceleratory factors were found in the blood of the unstressed animals *except* at those times when the same factors were peaking in the blood of the stressed animals—just after new and full moon. The points on the lunar calendar when the activity of the blood of stressed animals rose coincided invariably with the point in time when the activity of the blood of unstressed animals dropped to zero.

Why were the harried roaches' hearts aflutter just after new and full moon? Why are the hearts of mice and men under stress also likely to beat faster at such times? Mr. Rounds speculates that a change in the Earth's electromagnetic field, caused by the Moon, is the responsible factor. The pull of the Moon might trigger a faster heartbeat in an individual already under a strain. Quite as relevant is the completely opposite reaction of unstressed animals. Is stress a key factor in determining whether we enter a state of positive or negative receptivity to lunar influence?

In any event, there now is evidence that the lunar cycle can trigger excitement in the hearts of animals, including the human animal, already set on edge by circumstances. Changes in water flow among the fluid compartments of the body and an increase in the heart rate would certainly make a person more likely to lose his cool in an emotional situation.

Harry Rounds's data showed peaks that were lagged just past new and full moon. In fact, those lags of two to three days compare well with the lags in our Cuyahoga County homicide study and with Dr. Brown's hamster activity study. Wichita, where Dr. Rounds worked, is at a much higher latitude than Dade County, though not as far north as Cleveland or Evanston. The geographical factor again showed itself to be a key element in lunar-timed events.

Two additional studies extended evidence of the lunar effect on the blood stream and provided indirect support for the biological tides hypothesis. Dr. Edson Andrews, of Tallahassee, learned that the incidence of excessive bleeding during tonsil and adenoid operations was greatly increased around the times of new and full moon. Dr. Andrews had heard from his hospital staff that more bleeding occurred at full moon and decided to test this common belief. He used the records of over a thousand cases to investigate the lunar periodicity. He concluded that patients were more likely to bleed excessively at full *and* new moon. When he expanded his study, Dr. Andrews found that more bleeding-ulcer attacks occurred at full moon as well. He had discovered and recorded a red biotide. His findings were replicated by Dr. W. P. Rhyne, of Albany, Georgia, and by Ralph W. Morris, a pharmacology professor at the University of Illinois School of Medicine.

Dr. Andrews was so convinced by the results of his investigations that he says he threatened "to become a witch doctor and operate on dark nights only, saving the moonlit nights for romance."

Folk tradition has it that castration of farm animals is dangerous during full moon because of excessive bleeding. In ancient Jewish tradition, the practice of bleeding patients for medical purposes was limited according to the phases of the moon. The Talmud states that certain days on the Hebrew lunar calendar that match roughly with new and full moon are too dangerous for bloodletting.

Careful observers living by the lunar calendar have noticed certain periodic effects of great importance in daily life. Because we now live exclusively by solar timing, we have been ignoring potentially life-saving information.

Confronted with the evidence of increased bleeding in surgical patients at times of new and full moon, we must ask ourselves, What is the mechanism responsible? My answer to this difficult question is speculative and in accord with my theory of biological tides. Blood vessels and cells have semipermeable membranes. Body water passes freely back and forth among the fluid compartments. In a biological high-tide situation, bloating and tissue tension make it necessary to establish a new equilibrium among the fluid compartments of the body. It is likely that water under increased pressure will be reabsorbed into the circulatory system, resulting in increased blood volume and higher blood pressure. People with bleeding ulcers or with weak, delicate blood vessels may then tend to have spontaneous hemorrhages because of the sudden rise in volume and pressure. Patients who undergo surgery may have more of a bleeding tendency at these times.

It is evident, then, that there is considerable support for the biological tides hypothesis from a wide variety of sources—from folk traditions to the more sophisticated techniques of modern biological experimentation. We have speculated on two possible ways in which the Moon could affect basic rhythms—through a direct pull on the fluids in our bodies and through disturbances in the electromagnetic field, picked up somehow by our nervous system.

It has been demonstrated that at new and full moon our moods will change. If we are tense, our pulse quickens. If we are cut, we bleed more freely. The latter two effects were seen in ordinary individuals, and with wide statistical bases.

Our Dade County studies concerned people who were definitely *not* acting in an ordinary manner. They acted in a violent, aggressive manner. The same lunar periodicity showed up in their behavior. It is my contention that these people were unable to handle the added periodic stress imposed by a biological high tide.

We are at the beginnings of a new and holistic view of the Universe—a system in which each part and every organism resonates with the cycles of the cosmos. But how new is this view? Wasn't this what the old astrologers and early astronomers meant by the music of the spheres? Scientists are rediscovering what the ancients knew: the harmony of the heavens does not preclude a good deal of celestial dissonance as well.

6

The Moon and the Geophysical Environment

You can look at the distortion of the Moon as a result of the Earth pulling on the Moon, because it is kind of pear-shaped, toward the Earth. So you know, if the Earth can do that to the physical Moon, you can imagine what it's doing to the Earth. It probably affects human beings, too.

—James Irwin, astronaut
and Moon-walker

It is not generally realized that the gravitational pull of the Earth causes moonquakes or that the Moon has an effect on earthquakes. Tides of the oceans are accompanied by tidal stresses on the crust of the Earth. When it is at zenith over a particular area of the Earth, the Moon pulls that area's water and land outward in "a tidal bulge." At the same time there is a corresponding bulge on the opposite side of the Earth. This occurs because those areas of the planet nearest the Moon are subjected to a stronger gravitational pull. The surface of the ocean areas facing the Moon is pulled closer to the Moon than is the bottom of those areas, and directly on the other side of Earth the bottom of the ocean areas is pulled closer to the

Moon than is the surface of *those* areas. There are thus surface bulges in both such areas, and where there are surface bulges there are high tides. Halfway between the bulges around the globe are the regions of low tides, which provide the water for the high tides of the bulge areas. The planet rotates beneath the tidal bulges, and there are two high tides and two low tides daily.

THE TIDES

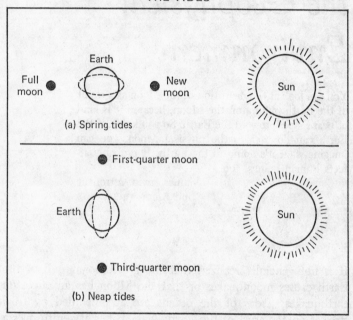

(a) Spring tides occur when the tidal forces of the Moon and the Sun work together to produce the high and low tides. (b) Neap tides occur when the high tides caused by the Sun are superimposed on the low tides caused by the Moon.

The Moon distorts the Earth as if it were a rubber ball. The ocean absorbs the force by sloshing around, but the land is more rigid; it cannot deform as easily and as quickly as water. The strain on the Earth's crust acts as a force in the triggering of earthquakes. (When the Moon is directly overhead, the North American continent may rise up to half a foot.)

In their fascinating book, *The Jupiter Effect*, John R. Gribbin and Stephen H. Plagemann reported the significance of tides as triggers of earthquakes: "Careful statistical analysis of large masses of earthquake data shows that the effect does occur but is not of overwhelming importance in itself. What is of the greatest importance, however, is that this work provides firm confirmation that earthquakes can be triggered by events occurring beyond the Earth." Plagemann and Gribbin are concerned that the grand alignment of the planets in 1982 may trigger another great California earthquake. (Of course, this coincidence of cosmic cycles could also lead to disruptions in human behavior.) Their documented work provides further support for the cosmic perspective. Our geophysical environment—even the ground beneath our feet—is being affected constantly by celestial events!

One aspect of the Earth's environment that is of daily concern is the weather. A connection between the weather and the Moon had long been suspected, but only in the last few decades has it been conclusively demonstrated. In 1962 three meteorologists—D. A. Bradley, M. A. Woodbury, and G. W. Brier —published a paper in *Science* magazine showing a definite lunar periodicity in heavy precipitation. To organize their data, based on weather records from 1900 to 1949, they divided the lunar month according to the synodic decimal scale. They found a "marked tendency for extreme precipitation in North America to be recorded near the middle of the first and third weeks of the synodic month, especially on the third to fifth days after the configurations of both new and full moon." Again there is evidence of a lunar periodicity and of precipitation peaks lagged just after new and full moon.

In the same issue of *Science*, E. E. Adderley and E. G. Bowen, researchers in Sydney, Australia, reported similar results derived from the records of fifty weather stations in New Zealand over a twenty-five-year period. These authors believe lunar periodicity cannot be used yet to predict the weather. They note it cannot be explained how the effect is caused, and they speculate the Moon's pull might alter the amount of meteoritic dust entering our atmosphere and affecting storm generation.

Not surprisingly, Messrs. Adderley and Bowen met with scientific prejudice against Moon-related effects. For a time they had withheld their results from publication. "To suggest a lunar effect on rainfall would simply not have met with the right response," they wrote. They were right in some quarters.

Supporting evidence did come in. In 1972, T. H. Carpenter, R. L. Holle, and J. J. Fernandez-Partagas published in *Monthly Weather Review* a study that demonstrated a *significant* relationship between lunar phases and the formation of tropical storms and hurricanes. Their study differed from the previous ones, which had measured the greatest rainfalls. The new study was concerned with the most violent storms. Using records of such storms in the North Atlantic and the Pacific Northwest, the researchers found that about 20 per cent more hurricanes and typhoons formed near new and full moon than near the quarters during a seventy-eight-year period, with a stronger peak at new moon than at full moon.

Everyone knows that the weather affects moods. We are more likely to be morose on a rainy day and feel sprightly on a blue-sky, sunny, "high-pressure" day. Is there a connection between lunar influence on the weather and lunar periodicity in human behavior? We know so little about the causes and the processes behind the observed effects that it is a tantalizing task to try to make causal connections from these fascinating observations.

Two researchers, Sheldon Geller and Herbert Shannon, of Toronto, believe they have discovered a link between the Moon's influence on the weather and bizarre human behavior. They have dubbed it "the Transylvania effect." It is their contention that the full moon in summer "is associated with uncomfortably hot and humid weather that interferes with normal sleep patterns and rapid eye movement sleep. The resulting deprivation of rapid eye movement has been linked to psychotic states."

Widespread precipitation and storm formation concentrated around full and new moon lend support to the hypothetical link connecting Moon, weather, and mood. Here may be another indirect effect of the Moon on normal biological rhythms —in this case, on the rhythms of rapid eye movement sleep (REM). There seems to be no lack of routes for the Moon's influence.

John Cejka, of Cyclomatic Engineering, believes enough is *now* known about the Moon's role in making weather to engage in long-term predictions. There is a correlation between tidal movements of the oceans and movement of the air masses (the weather). This relationship, Mr. Cejka writes, has a built-in fifteen-month time lag due to the difference in density between water and air. The long delay limits the accuracy of his predictions.

Lags and uncertainty characterize the Moon's influence. The complexity of time lags and harmonic relationships in weather, the strange lags in murder statistics, the reversed swings in electrical potential in experimental subjects, positive and negative receptivity—all point to the importance of intermediate processes, trigger effects, and unknown connecting principles in understanding the Moon's influence.

As I noted earlier, there are lags in lunar-timed events varying with latitude. The farther north of the equator the observed event, the longer the lag. One can only speculate why the influence of the Moon on behavior varies in time with the latitude of the location—but it is a fascinating realm for speculation.

Because the Earth is curved, a location in the north would be, to varying degrees, farther from the Moon than a location nearer to the equator. One would expect the Moon's gravitational force to be somewhat diminished over the somewhat greater distance, but how could that account for the delay in the timing of its effects? If the Moon's tidal pull is considered to have a *cumulative* effect, the delay begins to make sense. In northern latitudes, more days of spring tides would be required to create, cumulatively, a state of high nervous irritability. The effects of tidal forces on biological rhythms would accumulate to a critical level only when tidal forces were in decline.

We are already familiar with a cumulative lagged effect: the seasons. In the Northern Hemisphere the Sun is at its highest on the official first day of summer, usually June 21. In the months immediately following summer solstice, the weather gets hotter even though the Sun is retreating toward the equator and getting lower in the sky. This is because there is a cumulative effect in the warming and cooling of the atmosphere. It takes months of higher sun to warm the air enough to produce summer weather. It also takes many months of lowering sun for the accumulated warm air to cool. We are still determining if the *Moon's* effects are similarly cumulative.

Lags in the timing of periodic events, varying with latitude, are a commonly observed phenomenon. One researcher, Leonard W. Wing, has made many studies of the timing of natural cycles, such as migration of birds and fluctuations of rodent populations. The timing of many natural cycles tends to occur later and later nearer to the equator. Unfortunately for our argument, these lags vary conversely with observed lunar-timed lags. Mr. Wing has termed the tendency for timing that varies with location "latitudinal passage."

One instance of "latitudinal passage" in which timing *is* delayed poleward occurs in the fluctuations of cosmic rays: "As cosmic rays are of extraterrestrial origin," he notes, "it would seem they reflect ionosphere behavior." The pull of the Moon is of extraterrestrial origin, of course. It would seem the lags discovered in lunar-timed behavior are another example of latitudinal passage delayed in a northern direction. We know the ionosphere modifies cosmic forces such as radiation and magnetic disturbances, and that the Earth's magnetic field protects us from cosmic radiation. Is the Moon's influence modified by these planetary protective systems?

The importance of the delay according to latitude cannot be overstressed. It is a mysterious effect that gives us some clue, if we could only understand it, to the workings of lunar power. Ignoring it leads only to dead-end results. Drs. Pokorny and Jachimczyk, in attempting to replicate our research, tested for a correlation among the homicide curves of Dade County, Cuyahoga County, and Harris County in Texas. They found no significant statistical results. Their failure could have been foretold from our research, which had indicated results would differ according to geographical location. By ignoring the significance of geographical location, they arrived at meaningless results. Space—as well as time—is a critical variable for geophysical research.

The significance of the geophysical environment in our lives is often overlooked. Marshall McLuhan points out environments simply are taken for granted. People are unconscious of their effects because the effects are omnipresent. For most of us, it would be a burden if we were constantly conscious of the magnetic fields that envelop us or of the position of the Sun or of the phase of the Moon.

Some scientists have made the planet's changing, fluctuating environment their life's work. They work in disciplines with names like heliobiology, meteorpsychiatry, and biometeorology. Even among those who investigate environmental effects, consciousness of environment is not constant.

Yet, life depends on subtle and pervasive effects. One striking example of the importance of the unseen environment is the spatial orientation of some animals—birds, fish, and snails, for instance. Studies of migration indicate these animals pick up subtle clues present in the electromagnetic environment of the Earth. They home in on electrical patterns and are able to find their way accurately over long distances. This innate ability has been described as the "biological compass."

Dr. Brown has performed laboratory experiments in which the directional sense, or biocompass, of worms was changed by simply rotating the experimental apparatus by 180 degrees, shifting the worm's orientation to the Earth's magnetic field. The experiment shows conclusively that an animal actually can sense the Earth's magnetic field and use it for direction-finding.

Subsequent work by Brown and his associates revealed that insects and higher animals were indeed responsive to rotating horizontal magnetic fields similar in intensity to those of the Earth's field. In addition, this responsiveness varied with the time of day and with the phase of the Moon. Studies by other researchers have indicated organisms such as honeybees, flies, robins, homing pigeons, and gerbils are influenced by magnetic fields no stronger than those of terrestrial geomagnetism. Homing birds appear able to identify migratory directions by means of magnetic fields in the environment and gravitational forces.

Homing pigeons can be thrown off course by the attachment of small bar magnets to their necks. They can be reversed in their flight pattern simply by reversing the direction of the ambient magnetic fields produced by minute magnetic coils attached to their heads. Dr. Brown concluded that a living organism possesses the capacity to employ a subtle geophysical field component such as magnetism to distinguish geographical directions; in other words, the organism possesses a reliable magnetic compass. The ability of living organisms to sense magnetic fields applies not only to direction finding. It seems also to serve a regulatory function.

Reversals of the magnetic field can result in 180-degree phase shifts in biological rhythms of several organisms. Hence, an alteration in the *spatial* vector of an organism's environment can result in changes in the *timing* of its biological rhythms.

There are many factors that influence the Earth's magnetic field and, in turn, organic life. One of the factors, to be sure, is the Moon. The Moon affects the magnetic field of the Earth in two ways. It has its own magnetic field, which, though weak, causes a rhythmic fluctuation in the Earth's magnetic field. There is also an indirect effect on the Earth's magnetic field. It is brought about by the Moon's influence on the weather, which, in turn, influences electromagnetic patterns.

A Soviet scientist speculates the Moon may be responsible for the disappearance of aircraft in the infamous Bermuda Triangle. A. I. Yelkin, of the Moscow Institute of Building Engineers, believes lunar-solar tides disturb the Earth's magnetic field beneath the ocean, causing aircraft compasses to give false readings—often with tragic results. Mr. Yelkin's plotting of plane disappearances in the Atlantic indicates they occur during full moon, new moon, or when the Moon is nearest Earth, a close encounter of all kinds.

Klaus-Peter Ossenkopp, of the University of Manitoba, has engaged in extensive research in the area of geophysical influences on human behavior. He has written that ". . . there seem to be definite lunar influences on weather formations, and these changes in meteorological conditions would seem to indicate a change in the ambient electrical and magnetic fields, which have been shown to influence the behavior of animals."

By now it should be evident that so-called constant laboratory conditions do not exist. Even if an organism is shielded in a thick lead container buried deep in the earth, it still is susceptible to penetration of subtle geophysical forces, such as gravity and electromagnetism. The total composite of information reaching organisms by way of pervasive forces possesses information relative to sunrise and sunset, moonrise and moonset, and the celestial longitude of the Earth on any and every day.

Organisms appear to be most intimately and continuously associated with their environment. The organism itself is a dynamic electromagnetic entity, with fields that are continuous with the fields of the environment. One of the pioneers in research on the dynamic balance of life and environment was the late William F. Petersen, professor of pathology at the University of Illinois School of Medicine, in Chicago. During a thirty-year period, from 1920 to 1950, he performed painstaking research that revealed human biological rhythms correlated with weather patterns. He found that both human biorhythms and weather changes could be traced to natural fluctuations in the solar cycle, the lunar cycle, and the star cycle.

Dr. Petersen worked at a time when medical science was indifferent to biometeorological research. He received virtually no financial support or encouragement. In one particularly important experiment, he studied the total biochemical and physiological responses of a set of identical triplets who were then sophomore medical students at the University of Illinois. The experiment yielded curves for every function studied. They were correlated with the major weather variables in the local environment. Changes in biorhythm curves were related to disturbances such as the passage of warm and cold fronts, movement of storms, and barometric pressure changes. Dr. Petersen was able to trace the occurrence of disease and psychological disturbances to changing meteorological conditions. He also was able to correlate atmospheric tides with the organic rhythmic changes in the triplets.

Dr. Petersen came to believe that the critical factor in the rhythmic effects in health and environment was the adequacy and integrity of the available oxygen in the atmosphere. He has not been supported in this assumption by any work that I know. Whether oxygen is a critical factor or not, his statistical correlations remain. They have been supported by subsequent research.

Dr. Petersen documented rhythms in the general population by investigating numbers of births, deaths, suicides, psychotic episodes, sex ratios of newborn infants, and sex ratios of the dead. The death curve was related to the admission of psychotic patients, births, and weather variables. Atmospheric tides were correlated with rhythms in births, deaths, and psychotic disturbances in the general population. Dr. Petersen concluded that the population was reacting en masse to cosmic forces, the weather being the intermediate link to the cosmos.

Most of Dr. Petersen's work involved correlating individual and population rhythms with the sunspot cycle. He was also interested in detecting lunar periodicities. The inspiration was the work of the Danish Nobel Laureate Svante Arrhenius, whose paper "Cosmic Influences on Physiological Phenomena," in 1898, marked the beginning of modern scientific studies on the cosmic effects of human organisms. Arrhenius demonstrated clear-cut lunar cycles evident in deaths, births, menstruation, epileptic attacks, and atmospheric electrical potential. Dr. Petersen was able to identify lunar rhythms in the following data samples: birth records in New York and Chicago, death records in New York and Chicago, the occurrence of scarlet fever and epilepsy, cardiovascular deaths, deaths due to tuberculosis, and suicides. He found that more male children were born immediately after full moon and that more female children were born at the time of new moon. Lunar trends were identified in weather variables.

Dr. Petersen concluded that "the periodic distortion of the Earth's crust and ocean surface with the changing position of the Moon is an impressive phenomenon and it is not impossible that such stresses might be reflected in the changing distribution of the fluid tensions in the organisms and might find clinical expression in functional alterations in susceptible organs and tissues so that they are out of balance. Normal individuals remain unconscious of the swing in the tide of their own bodily processes. It is only when amplification of the waves is too great or when the shock is too sudden that we have no time for proper equilibration and at such times we experience discomfort and disease."

Dr. Petersen was one of the founders of the now recognized field of biometeorology. He was one of the first modern researchers to put forward the concept of human beings as cosmic resonators, although he emphasized that the viewpoint was in fact quite old. He regarded much of his work as rediscovery. He stressed the idea of the human organism as a being dependent on and integrated with its total environment, which, in turn, was dependent on the rhythms of the Universe. Dr. Petersen used the term *cosmobiology*. It is the ideal term to describe his total outlook.

The Russians have developed the cosmobiological science of heliobiology. In essence, it is the study of the Sun and its influence on terrestrial life. A number of studies show correlations between sunspot cycles and *sun*dry events such as accidents, epidemics of plague, crop yields, viral illnesses, and heart disease. The sunspot cycle of eleven years has been correlated with such events by researchers all over the world.

In 1915 the Russian scientist Chizhevskii initiated a systematic investigation of the relationship between several biological phenomena and cosmic variables. He found parallels among changes in solar activity and colloidal-electric changes in the blood, lymph, and cell protoplasm of animals, and the growth of bacteria cultures. Later it was discovered that diphtheria bacilli became less toxic in the years of maximum solar activity and more like closely related bacteria of a harmless nature.

In 1935 Japanese scientists observed a relationship between the rate of clotting of human blood and solar activity. At the time of passage of spots across the central meridian of the Sun, the index of the blood-clotting rate more than doubled. This effect was correlated with the rotation period of the Sun (twenty-seven days) and with the eleven-year cycle of solar activity. In 1958 it was found from a study of almost fifteen thousand cases that the total white blood cell count was reduced in periods of solar activity, although the number of lymphocytes increased. A characteristic geographical feature of this effect was noted. It was most pronounced in the polar regions and practically absent in the equatorial latitudes. (Because of the structure of the Earth's magnetic field, the Van Allen belts of radiation have holes at the poles: they are like a big doughnut wrapped around the globe. The poles are thus less protected from cosmic radiation.)

It has long been known that the size of populations of a great variety of organisms increases in cycles, with a period of close to eleven years. This cyclicity has been observed in the growth of marine algae and coral colonies and in the reproduction of fish, insects, and several mammals. A correlation between increased sunspot activity and the incidence of various diseases has been observed repeatedly during the past century. Chizhevskii, in the late 1930s, showed that outbreaks of plague, cholera, influenza, diphtheria, and other infectious diseases coincided with periods of increased solar activity.

Solar flares associated with sunspot activity cause magnetic storms on Earth. These perturbations are held responsible for coincidental effects on life. Dr. Alexander Dubrov, a heliobiologist in the Soviet Union, believes there is a definite connection between changes in the Earth's magnetic field and cardiovascular disease. The connection, he says, lies in the geomagnetic field's ability to cause "acute changes in the permeability of the blood vessels." According to Dr. Dubrov, these acute changes are brought about through influence on a "magical" substance—water: "Apparently the control is connected with changes occurring in the water molecules of the membranes."

This biological discovery is of tremendous importance. Membranes mediate so many vital processes in our bodies that we could elaborate endlessly on the implications of this cosmic influence! Dr. Dubrov concluded that ". . . the geomagnetic field, controlling permeability, has a decisive influence on all the processes that occur in living cells, in the organism, and ultimately in the entire biosphere."

His discovery struck me like a lightning bolt. I had based my theory of biological tides on the exchange of water among the body's fluid compartments, an exchange controlled by membranes. If the permeability of the membranes is controlled by the ambient magnetic field, and if the Moon influences the magnetic field of the Earth, a possible chain of causality has been established between the Moon and fluid balances in the body. If these insights prove out, Dr. Dubrov will have provided a link in the theory of biological tides. His work affects many areas of the biological sciences. His discovery of an effect of the geomagnetic field on heredity is of great potential importance as well.

As we have seen, Dr. Dubrov's results led him to suspect magnetic fields effect changes in water molecules. Water, so seemingly simple a substance, can undergo baffling changes. The inconsistent nature of water intrigued the late Professor Giorgio Piccardi, who was director of the Institute for Physical Chemistry, in Florence, Italy. He became involved with the question when he was asked to devise a way of cleaning lime deposits from industrial boilers. Professor Piccardi treated, or "activated," water, then used it to dissolve the deposits. His process worked well, but inconsistently. Used from time to time and from place to place, it displayed varying degrees of efficiency. Wondering why variation occurred, Professor Piccardi decided to study the properties of water further.

In his laboratory he set up test tubes of distilled water. Half of the tubes were left open to the environment and the other half were screened with metallic grids. To all the tubes he added the chemical compound bismuth oxychloride, which formed a colloid precipitate in the water.

Professor Piccardi found the rate of precipitation in the unscreened tubes differed significantly from that of the screened tubes. Over many years, thrice-daily measurements of the precipitation rates of colloids were made under different experimental conditions using both purified water and "activated" water. The rate of precipitation of the colloids was found to vary with solar activity, moon phases, cosmic radiations, movement of the planets in relation to the Earth, and other cosmic events. The results of his screening experiments led Professor Piccardi to deduce that variations in the precipitation rates were mediated by the Earth's magnetic field. Like Dr. Dubrov, he found the magnetic field was the medium through which cosmic events exerted an influence on the physical properties of water. It was natural then to surmise that if water could assume different physical states, depending on the natural variations of ambient geomagnetic fields, water in colloid solutions (which compose the bulk of most plant and animal organisms) would also be affected by cosmic occurrences.

Cosmic perturbations caused dramatic changes in the precipitation rates in the Piccardi experiments. The body fluids of animals and humans are doubtless affected in a similar fashion—a discovery of great importance to the biological-tides theory.

A change in the physical properties of body fluids could lead to changes in the retention of fluids, changes in the speed with which fluids pass through membranes, changes in blood pressure, and changes in heart output. The ability of cells to take up water, the electrical conduction among body tissues, and nerve firing would be affected as well. The "Piccardi effect" on water suggests the possibility of an effect on body processes by cosmic events through the medium of electromagnetic fields.

Professor Piccardi detected a difference in the physical properties of water and colloidal solutions at different geographical latitudes. He attributed the difference to the varying intensity of the Earth's magnetic field at various latitudes. The observation may prove important in the clarification of the phenomenon of latitudinal passage.

The work of Dr. Siegnot Lang, of the University of Saarland, in Germany, validates the concepts of Professor Piccardi. Dr. Lang was interested in determining the effects of electrostatic fields on the physiology and behavior of rats and mice. Animals placed in a Faraday cage, which effectively shields out 99 per cent of terrestrial electrostatic forces, showed dramatic increases in fluid retention and body weight, with significantly altered physiological and behavioral functioning. Hormonal regulation, body chemical processes, and neuromuscular irritability were markedly altered during observation. Faraday-cage shielding represented man-made interference with the pervasive electromagnetic fields surrounding the animals. In effect, Dr. Lang created a cosmic perturbation in the laboratory.

The alteration in the physical state of the animals' body fluids resulted in the dramatic changes in fluid and electrolyte metabolism previously noted. By causing an interference in the natural electromagnetic forces of the environment, a Piccardi effect and subsequently an artificial biological high tide were created in the animals.

Research into the influence of the geophysical environment upon behavior has increased of late, but we still are only skimming the surface. Most studies develop more questions than they answer. This is typical of a field in its early stages of growth. In 1963, Dr. Robert Becker charted a correlation between mental-hospital admissions and solar storms (or changes in the sunspot cycle). He later reported a correlation of geomagnetic disturbances with disturbed behavior of patients on psychiatric wards, and reported the effect of magnetic fields on the reaction time and performance of human subjects. His findings indicated modulated magnetic fields can significantly affect reaction time and, hence, affect performance of tasks by humans. Furthermore, the behavior of psychiatric patients was correlated with cosmic ray activity. (Cosmic rays provide a quantifiable measure that is related to the geomagnetic activity of the environment. The magnetic field of the Earth helps to protect us from dangerous cosmic radiation; the amount of cosmic rays that reach the Earth indicates the relative strength of the Earth's magnetic field.)

In 1968, R. Bokonjic and N. Zec of the medical faculty of Sarajevo University, in Yugoslavia, demonstrated a correlation between strokes and fluctuations in atmospheric pressure, temperature, and humidity.

Dr. Floyd Dunn, professor of electrical engineering and biophysics at the University of Illinois, and John Green of Bell Labs correlated the incidence of automobile accidents with high rates of infrasound. Infrasound is extremely low frequency sound. It is so low we cannot hear it. Although it is below our threshold of hearing, infrasound affects us in many ways. In moderate intensities, it makes us uncomfortable and agitated. The effect is thought to be responsible for the increase in accidents in the Dunn-Green study. Infrasound of great intensity can kill by disintegrating living tissues. Luckily, naturally produced infrasound does not reach this level. In the natural world, infrasound is a subtle, pervasive effect. It is thought to be generated by magnetic storms caused by solar disturbances.

Dr. Felix Gad Sulman has shown that periods of hot, dry wind, known variously as sharav, föhn, sirocco, or harmattan, are associated with marked increases in restlessness and irritability among people and animals. In areas that are flat and surrounded on three sides by mountain ranges, occurrences of this "ill wind" are correlated with increased suffering and death among the physically and emotionally ill. Apparently, the ionic balance of the atmosphere changes with the arrival of the ill wind. Excess positive ions from the atmosphere are trapped at ground level. According to the research of both Dr. Sulman, who works in Israel, and Dr. Albert P. Krueger, of the University of California, Berkeley, the effect on humans of excess positive ions is to bring about an increase in serotonin, one of the body's neurotransmitter substances. Excessive serotonin can make a person nervous, irritable, and depressed. If an individual is ill already or is predisposed toward emotional outbursts, the effect can have serious consequences. (Conversely, an excess of negative ions brings on a feeling of well-being. Plunging water generates negative ions. This is one reason so many people enjoy taking lengthy showers and being around great waterfalls like Niagara.)

People in areas visited by ill winds commonly explain bad moods and strange behavior as being caused by "the wind." When the ill wind is blowing, some surgeons will not operate, except in an emergency, believing that such times are particularly inauspicious. The approach of a low-pressure front also brings about positive ion concentrations. Just about everyone experiences a mood change when a storm is brewing.

Not only does the weather alter the ion balance in the atmosphere, it produces, as previously noted, electromagnetic disturbances. These take the form of extremely low frequency (ELF) electromagnetic fields. In their study on self-inflicted injuries and the lunar cycle, Klaus-Peter and Margitta Ossenkopp speculated that electromagnetic disturbances, indirectly linked to the lunar cycle, had an effect on human behavior. Finding a significant correlation between the lunar cycle and self-inflicted injuries in females, they asked themselves if the female hormonal system was particularly sensitive to this type of ELF disturbance.

ELF waves could be produced by the Moon's interference with the magnetic field of the Earth as well as by the weather. The Ossenkopps wrote: "If ELF waves from weather disturbances, or perhaps even the energetic particles released from the Earth's magnetic tail during certain parts of the lunar cycle, shifted in some manner the hormone balance of females to one similar in manner to menstruation onset, this might explain the distribution for females found in this study."

The Ossenkopps' speculation about the influence of the Moon on the magnetic "tail" of the Earth is fascinating. It is during the phase of full moon that showers of energetic particles may be released: "At full moon, the Moon traverses the tail and is in it for about four days." Scientists have suggested this phenomenon might influence the frequency of such widely varied events as self-destructive behavior and thunderstorms.

Meteorologists have speculated that the Moon's pull changes the amount of meteoritic dust entering the atmosphere. Indeed, the Moon can interfere with the patterns of other cosmic influences on our planet. If, for example, the Moon obstructs radiation coming from the Sun (solar wind), cosmic showers could be jammed for a brief time. It is probably a "small effect," but we should not write it off. In the course of research, we have seen how seemingly small effects can be of enormous importance.

Dr. Robert Becker (whose work correlating mental-hospital admissions with the sunspot cycle has been mentioned) pioneered in the area of extremely subtle electromagnetic effects. Dr. Becker is professor of orthopedic surgery at Syracuse University, as well as Medical Investigator at the Syracuse VA Hospital. In the 1950s he became interested in Burr's early findings on the difference in electric potential between whole and injured tissue, and began a thorough study of the electrical properties of wound-healing. Early on in his work, he identified electrical changes at the site of fresh tissue injuries. He called these changes "the current of injury" and subsequently showed these electrical changes were instrumental in the healing and regenerative processes of injured tissues. It is possible that the electrical difference between injured and healthy tissues is what has been recorded by Kirlian photography and labeled as energy fields.

Dr. Becker established that nerve tissue produced minute direct-current electricity. He was able to trace the production of minicurrents both to the nerve sheath cells that surround all peripheral nerves and to the glial and syncytial cells that surround and support the nerves in the central nervous system (the brain and spinal cord). These electric currents exist independently of the action-potential system that conducts nerve impulses along the nerve fibers themselves.

Dr. Becker believes that the direct-current electrical system, from an evolutionary standpoint, antedates the action-potential system and may, in fact, be the precursor of that system. The mapping of electrical potential systems of entire organisms has shown that electrical lines of force align themselves in a pattern identical with the neural structures of the organism.

As a result of his studies, Dr. Becker evolved a theory of the development and function of the nervous system. Its electrical conduction, he believes, is a hybrid process. On the one hand, direct-current conduction accounts for the growth and development of the nervous system, the organ system, and the cell processes of the organism itself. This direct-current system has properties resembling those of an analog computer with solid-state semiconducting functions. On the other hand, there is the familiar action-potential system that conducts nerve impulses along the nerve fibers. Its functioning resembles that of a digital computer.

Dr. Becker found that the two parts of the hybrid system are connected at "amplifier nodes" located at intervals along the neural network. Through these amplifier sites, changes in the direct-current system can amplify or modulate the traveling of action potentials along the nerve fibers. These amplifier nodes are identical to the well-mapped acupuncture points described in ancient and modern literature.

It is via this direct-current system, Dr. Becker suggests, that we "sense" changes in the electromagnetic environment: "This system is involved in the receipt of damage or injury stimuli— which we perceive as pain—and in the control of various growth processes of repair, including bone regeneration. Its nature renders it susceptible to perturbation by electrical and magnetic fields. It is proposed that the direct-current system furnishes the linkage mechanism between biological cycles and geomagnetic cycles," which may interact at the proposed amplifier nodes.

Of course, we are not conscious of the environment's effects on this system. This is why its functions have presented us with mysteries that we could not explain. If Dr. Becker's theory is correct, we can understand more clearly how the biological compass operates, how magnetic storms associated with sunspots disturb our health and behavior, and how the Moon may be linked to the nervous system. According to Dr. Becker: "The effect of the Moon on living things is a 'second-order phenomenon' due to the Moon's position in the Earth-Moon-Sun geophysical system, producing cyclic alterations in the Earth's magnetic field."

Dr. Becker's work, in the main, has been in the area of wound-healing and bone-fracture regeneration. He has experimented with the use of tiny electric currents, generating low-intensity electromagnetic fields as therapeutic treatment. This technique, he believes, is a simple encouragement of the natural healing process, which makes use of similar biologically produced currents. Though his technique has met with substantial success, Dr. Becker cautions against its premature widespread use. He emphasizes that all the effects are not known.

When Dr. Becker first published his findings in the 1950s, his fellow scientists scoffed. But since then, there has been growing acceptance of his views by the scientific community.

Similarly, there has also been resistance to acceptance of the findings of Soviet scientists in related fields, partly because of lingering cold-war attitudes and partly because the Russians have done extensive work in areas that most American scientists deem unworthy of attention. Only now are we starting to realize that the Russians have a great deal of accumulated knowledge about the geophysical environment.

Russian scientists learned that the biological function of plants, animals, and human beings is affected by experimentally produced electromagnetic fields (herein referred to as "emfs") about as intense as those surrounding the Earth. The processes of growth and development, among others, are affected by fluctuations in ambient electromagnetic fields. These fields also have an organizing effect on the cellular structure of plants and animals, and they may play a vital role in the evolution of life.*

The effects produced by low-intensity emfs are independent of the thermal and destructive effects produced by high-intensity electromagnetic radiations, like X-rays, gamma rays, and super high-frequency radio waves. Many of the effects that Dr. Presman has documented are produced by weak fields close to the level of natural electromagnetic fields. He postulates that such fields are conveyors of information to the organism, among separate organisms, and even within our bodies.

Living beings are constantly buffeted by noisy fluctuations in the natural emfs that surround us. This emf "noise" is caused by such phenomena as sunspot and solar-flare activity or coincidence of planetary cycles. It has been established that the periodic variations of the natural environmental emfs have a regulating effect on vital functions—for example, on the rhythms of the main physiological processes, on the ability of animals to orient themselves in space, and on multiplication in populations. In living organisms, the systems for receiving emf-

* Much Russian-developed data were not available to us until recently. Aleksandr S. Presman's book *Electromagnetic Fields and Life*, translated at the start of this decade, provided much of what we now know about work in this area in Russian laboratories.

transmitted information are protected from natural electromagnetic interference. In pathological states, spontaneous perturbations of emfs (e.g., solar flares and lightning discharges) upset the regulation of physiological processes. There is evidence that, during the course of their evolution, living organisms have used emfs to obtain information about changes in the natural environment. By means of emfs, information can be transmitted through any medium inhabited by living organisms and under any meteorological conditions.

Dr. Presman's concept of the regulatory effect of natural emfs complements Frank Brown's extrinsic timing theory of biological rhythms as well as my biological tides theory. Exactly how information is transmitted by emfs is not known. It may be possible that disturbances in physiology are brought about by direct action of the emfs on the various parts of the nervous system. (This possibility supports Dr. Becker's idea of amplifier node receptor sites in the nervous system.) Dr. Presman believes that the central nervous system, the cerebral cortex, and the hypothalamus appear to be most sensitive to emfs.

Recently developed sophisticated equipment, first used in the space program, has been employed to detect minute emfs formed around active nerves, muscles, and the hearts of animals. The very fact of the detection of an emf around isolated cells and organs and around entire organisms indicates the possibility of exchange of information among animals by means of emfs.

Dr. Presman lists four possible types of bio-information transfer between and among animal populations. The first type of bio-information transfer is responsible for the rapid coordination of the activity of an individual in a group or community of animals. Such connections presumably underlie simultaneous changes in direction of birds in a flock or of fish in schools, and the rapid coordination of certain insects (acting as if they had "one mind"). A connection is made over relatively short distances and can be effected by weak signals carrying a small amount of information. The second type of transfer is a relatively slow signal transmission. It might be regarded as the explanation for the baffling ability of many animals to find their way to one another over a distance. The third type of bio-information transfer is the slow exchange of information by emfs among individuals of one population or one species. The fourth type involves the behavior and development of animals in groups and communities in interaction with environmental emfs. Though the Presman models of bio-information transfer remain vague, they may constitute a framework for the understanding of so-called extrasensory communication.

The work of many different researchers is pointing to the existence of an unseen but vitally important electromagnetic Universe. Through emfs, we pick up biological timing cues and communicate and regulate our living systems. Through emfs, we stay in tune with the cosmos.

The American researchers Stanley Krippner and Sally Ann Drucker have suggested that the technique of Kirlian photography can make visible the electromagnetic patterns of life. Kirlian photography, or electrophotography, involves photographing the radiation emitted by objects placed between electrodes creating a high-intensity field. Many Kirlian photographs show light emanating from well-mapped acupuncture points, which Dr. Becker believes may be the locations of amplifier nodes in the nervous system. Kirlian photographs, ancient and modern mapping of acupuncture points, and the latest theory of hybrid data transmission of the nervous system all appear to define the same phenomenon from different approaches and through different techniques. A pattern is coming together. We are beginning to *perceive* our connection with the cosmos.

Professor Presman demonstrated the sensitivity of organisms' emfs to magnetic disturbances. Confirmation should be demonstrated by electrophotography if it is indeed recording life emfs. According to Krippner and Drucker, Victor Adamenko related (in 1970) the changes in the fields of objects photographed by Semyon and Valentina Kirlian to variations in the Earth's electrical fields.

Researchers in many countries have found that Kirlian photography sheds light on another mysterious phenomenon: psychic healing. Electrophotography of the fingers of psychic healers shows dramatic changes when the healers "turn on" their healing power. Healers are practicing an ancient form of medicine. Is Dr. Becker's low-intensity emf therapy its modern equivalent? If so, scientists may have to change their notions about ancient medicine.

7
The Moon
in Evolution

Why did not the human line become extinct in the
depths of the pliocene? . . . We know that but for
a gift from the stars, but for accidental collision of
ray and gene, intelligence would have perished on
some forgotten African field.

—Robert Ardrey

The Moon was Earth's constant companion through the long
eons during which life evolved. Because our ancient precursors
are said to have emerged from the sea, it is reasonable to expect
they bore with them a special sensitivity to the tides. Evidence
of the Moon's influence on evolution constitutes a major argument in favor of the theory of biological tides.

The first to enunciate the modern scientific theory of the origin of life in the sea was Erasmus Darwin, the grandfather of
the famous Charles Darwin. In his biological epic, *The Temple
of Nature*, grandfather Darwin penned the immortal couplet

> Organic life beneath the shoreless waves,
> Was born and nurs'd in Ocean's pearly caves;

The idea that life emerged from the sea is not by any means
exclusively the product of scientific thought. Creation myths
allude to the ocean as the home of primordial life. Robert
Graves gives us this account of the Homeric creation myth:
"Some say that all gods and all living creatures originated in

the stream of Oceanus which girdles the world, and that
Tethys was the mother of all his children."

What led our ancestors to speculate that life originated in
the sea? ". . . and the gathering together of the waters He
called Seas, and God saw that it was good." Did they simply in-
tuit it? Is there racial memory of evolution that led to the for-
mulation of these myths?

Life's origin in the sea, now accepted as scientific fact, means
that the Moon, which governs the rhythm of the ocean tides,
was of great significance in evolution. The Moon's constant
rhythm accompanied each stage of the ongoing struggle of life
to assume more and more organized forms of existence.

In his classic *The Descent of Man*, Charles Darwin argued
that the Moon's influence was evidence for the evolution of
man from lower forms of life: "Man is subject, like other mammals, birds, and even insects, to that mysterious law that causes
certain normal processes, such as gestation, as well as the mat-
uration and duration of various diseases, to follow lunar periods." Darwin found the key to "that mysterious law" in the
stage of evolution at which life emerged from the sea and set
out upon the great adventure of the land animals that led to
the evolution of the human species. Emergence must have
taken many generations of adaptation. There developed animals with a form halfway between the true fishes and the later
amphibians. Such animals still exist; the South American
lungfish, for one example.

The traces of this ancient event are visible still in the development of the human embryo. In tracing the genealogy of
man, Charles Darwin noted: "At a still earlier period the pro-
genitors of man must have been aquatic in their habits; for
morphology plainly tells us that our lungs consist of a modified
swim-bladder, which once served as a float. The clefts on the
neck in the embryo of man show where the branchiae once
existed. In the lunar or weekly recurrent periods of some of our
functions, we apparently still retain traces of our primordial
birthplace, a shore washed by the tides."

At the critical stage in evolution, the rhythm of the tides
must have assumed great importance for these organisms. "The

inhabitants of the seashore," Darwin wrote, "must be greatly affected by the tides; animals living either about the mean high-water mark or about the mean low-water mark pass through a complete cycle of tidal changes in a fortnight. Consequently, their food supply will undergo marked changes week by week. The vital functions of such animals, living under these conditions for many generations, can hardly fail to run their course in regular weekly periods. Now it is a mysterious fact that in the higher terrestrial Vertebrata, as well as in other classes, many normal and abnormal processes have one or more whole weeks as their periods; this would be rendered intelligible if the Vertebrata are descended from an animal allied to the existing tidal Ascidians."

The daily tides must have provided occasional and welcome relief from the novel land environment in which these animals found themselves. Emergence from the sea, accompanied by the rhythmic beat of the tides, was a traumatic event as well as a triumphant adaptation. The pull of the tides is both nostalgic and anxiety-ridden. It recalls a time when a great drying up in one or another region compelled aquatic animals to seek a new means of existence.

This disaster—the dessication of the sea—was a great turning point in evolution in another sense as well. In his exploration of the evolution of sexuality, the psychiatrist Sandor Ferenczi suggested it was at just this point in evolution that aggression and genital sexuality were linked. A disciple of Freud, Ferenczi speculated that ". . . the warfare in connection with the earliest attempts at coitus was in reality a struggle for water, for moisture, and that in the sadistic component of the sex act this period of struggle achieves repetition, even though but symbolically and playfully and even though in such distant progeny of these earliest ancestors as the human species."

There are several primordial events occurring in a tidal environment accompanied by the rhythm of the Moon. Ferenczi's speculation on the origin of the battle of the sexes is particularly intriguing for us. He believes it began in a Moon-dominated environment. Such a view goes a long way toward explaining the connection of both sexual and aggressive drives with lunar periodicity. Rhythmic patterns of animal life are as much a product and a device of evolution as are eyes, fins, tails, and feet. But ingrained rhythms are not visible; in many cases they thus are overlooked. According to my colleague Frank Brown, "Living creatures are exposed to pervasive environmental periodisms wherever they are and at every stage of their life cycles, from reproductive cells through adult. One may speculate that very early in the course of the evolution of life, the recurring solar-day and lunar-day geophysical patterns served as a template upon which were fashioned biological sequences of causally related events."

Sun and Moon have left their imprints in the timing of life at its most basic levels. It is appropriate that the Sun's daily rhythm served as the basic template of regular timing, while the Moon's greatest influence came at a time of cataclysm, trauma, and exceptional creativity—certainly a transcendental moment in evolution. It is exactly as many cultural interpretations have it: the Sun is the rule, the Moon is the exception. In light of such evolutionary speculations, it becomes clearer how and why the symbolism of the Moon has a hold on the imagination—and why we connect the Moon with birth and trauma.

Even after life emerged from the sea and adapted to the land environment, the influence of the Moon continued to be felt. Nights of full moon, as was mentioned in chapter 1, were excellent times for nocturnal predators to strike. Our precursors, who were vegetarians, knew this. Those who took warning at full moon—and remained restless instead of sleeping—lived longer and left more progeny. This lunar pattern is ingrained in our genetic heritage.* The pull of the sea remains powerful. How painful and difficult it must have been for our ancestors to emerge from the sea can be seen by the example of the cetaceans (whales and porpoises), who returned to it.

Another possible lunar influence on evolution involves the

_____

* Further support for the evolution explanation of human restlessness during full moon has been supplied by Stephen H. Vessey, of the Department of Biology, Bowling Green State University. He reports that the night activity of rhesus monkeys is higher during full moon.

Moon's effect on the magnetic field of the Earth. Alexander Dubrov, the Soviet heliobiologist, found a relation between changes in chromosomes and changes in the magnetic field: "The seasonal course of the changes in this field coincides with the frequency of chromosome inversions in the wine fly. The facts obtained mean that evolutionary changes in organisms under natural conditions occur under the influence of the geomagnetic field."

Dr. Robert Becker agrees with Dr. Dubrov: The electromagnetic connection is important for evolution. Putting together the findings in this area, one can see that electromagnetic fluctuations are received directly by the nervous system and somehow affect the mutation of chromosomes. His own work with "life fields" has led Dr. Leonard Ravitz to a similar view. The nervous system, he suggests, evolved "as a result of dynamic forces imposed on cell groups by the total field pattern."

Sensing the importance of changing fields in relation to survival, some species have taken advantage of this strong evolutionary influence. Dr. Frank Brown noted that if a pair of bean seeds were placed in a dish of water, one seed would absorb the water while the other would not. When the compass orientation of the seeds was reversed, the water-uptake function was reversed. Under different geomagnetic orientation, different seeds will take up water. This flexible stance, Dr. Brown concluded, offers an evolutionary advantage: "The species is steadily provided with two possibilities for survival—plus and minus states—in its response to the natural fluctuations in its geoelectromagnetic environment." Geoelectromagnetic implies that the electrical, magnetic, and gravitational properties of our earthly realm should be treated as an interacting whole.

In extremely simple organisms, life adapted to the effects of constantly fluctuating magnetic fields by extending its options. In this adaptive response, species survival is valued over individual survival. How individual members of a species coordinate such a transcendent activity is one of the mysteries of evolution. Dr. Brown tells me the answer may lie in the interactions of electromagnetic fields, or emfs. Aleksandr Presman assumes a similar position. If the Brown-Presman view is correct, extrasensory perception may have evolved as a response of various species to environmental stress.

Earlier I suggested how the evolutionary paths of wolf and man crossed. The legend of lycanthropy persists because the wolf was important in the development of human hunting practices. Now it is evident that the pull of the Moon was a potent force in evolution long before that point in human development. Indeed, the sway of the Moon was established long before anything remotely resembling human beings or wolves had evolved.

The association of the pull of the Moon with sexual and aggressive drives, formed when animals first moved from the sea onto the land, would easily have been linked, eons later, with the aggressive behavior of the wolf. Lycanthropy turns out to be a highly symbolic human concept representing biological tendencies far older than either species involved.

A fascinating aside: The survival of the werewolf legend probably received added impetus from the seventeenth-century movement to confine lunatics and madmen in hospitals and prisons. Some manic depressives can be triggered into episodes of maniacal behavior more frequently at times of new or full moon. At such times, their metabolic processes, including those of beard growth, are accelerated. Perhaps it was the observation of rapid beard growth among hypermetabolic patients, coupled with their fits of wild behavior, that reinforced the werewolf legend in centuries past. Until relatively recent times, it was *dog*ma among many madhouse keepers that the insane had descended to a level of animality. This belief suited the werewolf metamorphosis. Until 1808 it was customary at certain phases of the Moon to chain and flog inmates of England's notorious Bedlam Hospital, "to prevent violence." We tend still to project onto the insane our own fears of irrationality. Mental hospital conditions are often deplorable even today. Once man is able to accept the "wolfish" tendencies within himself—his evolutionary heritage—he may no longer require a scapegoat, human or animal.

Striking parallels are sometimes discovered between an organism and the environment in which it lives. I mentioned earlier that lunar mythology and the poetic nature of lunar-inspired thought are distasteful to many rationalist minds. There is a duality of cosmic time: the solar calendar and the lunar calendar. This duality is paralleled by a traditional opposition of reason—the rational (Sun) and the irrational and intuitive (Moon). After all, the solar day is regular—rational. The lunar day is out of step with solar time—irrational. The duality has taken on immense cultural significance.

The parallel of the Sun-Moon configuration, so profusely represented in myth the world over, has a biological as well as a cultural basis. Not only are the influences of solar and lunar timing ingrained in our body rhythms, they are reflected in the structure of our brains and in the very nature of thought itself.

It is now well known that the two hemispheres of the brain are independent of each other to some degree. For example, in motor functions the right half of the brain controls the left side of the body and the left half controls the right side of the body. If the major nerve trunk joining the two hemispheres is cut, what is learned by one hemisphere is *not* transmitted to the other. We can literally be "of two minds" on any given matter!

The potential independence of the two brain halves normally expresses itself as an interdependence. Robert Ornstein, a research psychologist in California, believes the left and the right hemispheres of the brain are quite different from each other in the way they perceive and think. The left ["solar"] hemisphere perceives and processes information in a linear, rational manner. The right ["lunar"] hemisphere functions in gestalt perceptions and intuitions. As a consequence of this remarkable discovery, the symbolism of the Moon in culture and history may be better recognized by twentieth-century minds for what it is. Myths of the Moon and the Sun, and the complementarity they express, parallel the biological structure of the brain and the functions of thought. It is astonishing, but logical, that there should be such a striking parallel in the dual structure of the calendars, the dual structure of the brain, the duality of the dominant modes of thought and perception, and the duality of the Sun and Moon myths.

The dual structure of the brain confers a survival advantage because the functions of a damaged brain-half may be assumed by the other half. That is why the brain hemispheres can be made, by a surgical procedure, to operate independently of each other. Another pattern evolved according to this duality—a pattern essential to the evolutionary process of life. To have motion, there must exist opposition. The opposition of two legs produces walking and running. Similarly, I believe that the opposition of two ways of perceiving and thinking, "lunar" (intuitive) and "solar" (rational), is necessary for the optimal development of thought and creation.

Dr. Albert Rothenberg, a psychiatrist at Yale University, has evolved the concept of Janusian thinking as basic to the development of creative thought and process. Janusian thinking is the capacity of the creative mind to entertain opposite perceptions and concepts simultaneously and without resulting conflict. Such a mind can transpose apparent duality into unified aesthetic productions—art, music, poetry, scientific theory and, of course, creative styles of living and relating in a world filled with paradoxes and contradictions.

In our society, we have neglected lunar modes of thought and inspiration that our ancestors once valued. In overbalancing our emphasis on solar rationalism, we have slowed the evolution of creative thought.

8

The Moon
and Civilization

In some senses quite different from those imagined
by the earliest astrologers, we *are* connected up with
the Universe.

—Carl Sagan, *Other Worlds*

The ancients knew that they themselves were cosmic resonators.

Although they knew nothing of magnetic storms, sunspots,
and cosmic rays, and were unable to formulate the laws of universal gravitation, they assumed that everything in the Universe
—including themselves—was linked in some way to everything
else. In such an interconnected world, the Sun, Moon, and
planets were highly visible signs of the workings of the cosmos
—and their motions were predictable.

The early astrologers calculated the first successful predictions of the motions of celestial bodies. Given the state of
mathematics, these computations were extremely difficult.
(Today, in the Computer Age, we take computation for
granted.) Ancient astrologers and the pioneers of astronomy
needed great patience and discipline. (The calculation of the
orbit of Mars drove one sixteenth-century astronomer mad.)
Johannes Kepler, who formulated laws of planetary motion,
spent years on the Mars orbit, working out part of the basis for
modern calculus in the process.

Astrology was once a valid observational discipline. Twenty centuries ago the astrologers were, in fact, the astronomers. They observed heavenly bodies and correlated their positions with events on Earth. We have learned that many of their correlations, especially with regard to the positions of Earth, Moon, and Sun, were valid. Modern researchers in the areas of fertility and birth control have demonstrated the importance of the configuration of Earth, Moon, and Sun in birth and reproduction, as I noted previously.

With regard to correlations of events with the positions of the other planets, we must ask ourselves if their positions can cause a measurable effect on our biological system. Such an effect is indeed possible. If Venus and Mars happened to be in alignment with the Sun, Moon, and Earth, the gravitational effect would be summated even more heavily at that time.

In chapter 6 it was noted that Plagemann and Gribbin expect the coming grand alignment of the planets to exert a strong gravitational force upon our planet, triggering earthquakes. If these quakes are accompanied by a massive behavioral upheaval throughout the population, as would be likely, we will have an impressive validation of some of the tenets of the ancient astrologers and, of course, of our own theories. This is not present-day astrology with its complicated and symbolic system, but it *is* the kind of celestial knowledge the ancients sought and perceived through observation. It should be remembered that it is in combination and in configuration that planets have their effect on us. Taken singly, most of the planets are so distant that their pull is too weak to affect human behavior.

In ancient civilizations, people tried consciously to live in harmony with the Universe. Our contemporary civilization, on the other hand, is self-consciously at odds with the environment. In our greedy desire to produce more and more goods, we are in a death struggle with nature.

In its day, astrology served in a variety of cultures as an observational science and a system of interpretation. Character was interpreted by astrologers using the signs of the zodiac and the positions of the planets. Having a system of classification makes it easier to discuss matters as complicated and as variable as the human personality. Twentieth-century personality interpretation uses different sets of metaphors. Professionals in the psychological disciplines may use the system of Freudian metaphors—borrowed largely from myth—or the technical jargon of the behaviorists.

In the past, the practice of living in harmony with the Universe extended to *all* aspects of life, not just to the observation of the heavens. Examples of this way of life are still with us today. The Ananda Marga Yoga Society prescribes fasting days according to the lunar calendar—the fourth day before full moon and new moon. Gravitational stresses begin to accumulate at these times, and the Ananda Marga believe they imbalance the system. Fasting and meditation at such times are said to help keep or restore the balance. Here is an interesting opportunity for applied lunar knowledge. If studies were to prove that this method of dealing with cosmic stress works, it could be of benefit to those individuals who are delicately balanced and likely to be adversely affected by environmental stress periods. (See chapter 10 for more applications of lunar knowledge.)

There have been occasional scientific confirmations of the predictions of astrology. C. G. Jung, in a now-famous experiment, found that the positions of Sun and Moon at the birth of individuals predicted—in a significant number of cases—the astrological conjunctions formed by their eventual marriages. The French psychologist Michel Gauquelin found a correlation between the career choices of successful professionals and the planets dominating the sky at the time of their birth. He came to believe that a child's genetic makeup makes it more likely to be born when the solar system is in a particular configuration. In his book *The Cosmic Clocks*, Gauquelin writes that, in the womb, the embryo is protected from the usual time signs, such as the light of the Sun. However, gravity and electromagnetic fields will penetrate to the embryo's environment. The positions of the planets will thus influence the time of birth. If organisms "are placed outside the reach of the 'obvious' timegivers in the environment, they will instinctively find other patterns by which to regulate their biological rhythms, becoming most sensitive to the influence of 'subtle synchronizers' from space."

Gauquelin reasoned that the genetic trait that causes a child to be born at a certain planetary configuration is linked to the traits that determine the child's talents. This view would explain the striking correlations of planets and careers: "Very simply, the child's career depends on the genetic structure of his organism; at birth, the planetary clocks reveal this genetic factor in an unforeseen way. The successful professionals had certain elements in their genes that allowed their lives to develop naturally in a favored direction, inherited from their parents. Of course, this relationship does not apply only to celebrities, it applies to everyone. In the human species, the inherited tendency to be born at a given hour instead of another should to a certain extent be an indication of the individual's constitutional type."

The studies of Jung and Gauquelin seem to lend validity to some facets of astrology. Though a few serious and qualified workers are applying modern research methods to the study of astrology, I know of little in the way of valid findings in the scientific literature.

I am not a believer in, or a supporter of, astrology. It does, however, embody a tradition of ancient observations and correlations. We must remain open-minded and willing to examine any system that seems to have heuristic value. Scientific qualifications are not a license for smugness, nor do they legitimize prejudice or bias.

Can we apply our growing knowledge of lunar and cosmic effects to daily life? Can we learn to live more in harmony with cosmic cycles? Some scientists think so. J. E. Davidson of Sandia Laboratories, in Albuquerque, New Mexico, studied accident patterns and came up with some surprising conclusions. He found that accidents were influenced by many cosmic cycles, the phases of the Moon in particular: "Our data suggest the possibility of a heightened accident susceptibility (error, misjudgment, etc.) for people during the lunar phase similar to that in which they were born and for the lunar phase that is 180 degrees away from that in which they were born."

Such scientific testing appears to lend credence to my theory that an individual's biological rhythms are imprinted at the time of birth. This could be of use in the prevention of accidents. Mr. Davidson also found correlations of accidents with sunspot cycles and magnetic fluctuations. Though it seems obvious that more exploration of this sort of knowledge is needed, one still runs up against ever-present bias. Myth or legend-laden material appears to be anathema to a large number of scientists. Why this should still be so is problematical.

If a concept persists over many centuries in a relatively unchanged form, if it is handed down intact as myth or legend, we should take a serious look at it. We should examine the observational basis for it. We should attempt to document scientifically what is going on. In the past, legend has often proved to be more accurate than history. After all, history is well known to be biased toward the predilections of the historian who relates it—and the historical interests of the powers that employ him. The problem of observer bias is a thorn in the side of science, as well as of history.

Another bias that has impaired understanding of the Universe is the tendency to reckon time purely on a solar basis. The solar time bias reflects the basic imbalance of our age and our culture. In ancient society, the balance between Sun and Moon was a concept central to understanding the workings of the cosmos, civilization, religion, and personal well-being. The solar principle symbolized the rational, regular, dependable proceedings of life. The Sun was the giver of life and the bringer of light. Its cycle was simple and precise. The Moon, on the other hand, was out of synchronization with the Sun. The effect of the Sun on temperature and crops was obvious, while the effect of the Moon on ocean tides and on the tide in the affairs of men was subtle and difficult to work out. The Moon seemed irrational, as opposed to the consistent bright periodicity of the Sun. The Sun was the symbol of everything that was reasonable, regular, and dependable.

As our research has proven, and as all ancient civilizations knew, the Moon has a tendency to bring out the irrational in man—the unbalanced behavior in which even sober souls may indulge at times. Of course, the lunar realm does not comprise solely the wild and crazy side of life. Other mysterious but necessary qualities of life, like intuition and creativity, were linked by ancient societies to the Moon.

In the course of Western history, the solar view of the world won out. Instead of living with a balance of the rational and the intuitive, Western science chose to deny and repress the intuitive side of life. Mysterious effects that could not be explained by reason were, one after another, denounced as superstition, myth, and witchcraft. The witchcraft craze in Europe came about after the old religions and mythologies were driven underground by science and the Church.

The irrational, the intuitive, and the creative aspects of personality arise from the unconscious mind; these "lunar attributes" were valued as socially necessary by the ancients. When the myths, legends, and forms of Moon worship were officially repressed during the Dark Ages, they became a sort of social collective unconscious, liable to erupt only in such phenomena as mass panics and heresies. The old duality of Sun and Moon was reinterpreted as the opposition between God and the devil. What had once been conceived as a cosmic harmony was turned into a universal dissonance.

This dissonance showed up in my work when I tried to understand lunar influences in terms of rational, linear thought. A universal and unresolved problem in the study of natural environmental influences on human behavior is the basic unsuitability of statistics. No statistical test and no area of mathematics has been developed that is capable of capturing periodic and continually fluctuating variables. Statistics are static. They are only capable of providing a cross section of what is happening at any one point in time. The cross section is artificial at best. The method and the subject matter are not suited to one another.

We now have highly accurate scientific tools to work with, but they are of little use in dealing with certain problems. This is especially obvious in psychology, where every day we see the attempts of the rational to understand the emotional. Currently, rational science is better at manipulating behavior than at understanding the emotions and inspirations that are so important to well-being.

Opposition to and repression of the irrational is not unique to recent history. Although ancient civilizations used lunar calendars and often thought in terms of a balance of lunar and solar modes of thought, repression was already underway. In the myths of almost all cultures, the Moon is represented as a goddess and the Sun as a god. The opposition of Sun and Moon is as old as the battle of the sexes. Certainly, the worship of the Moon goddess began to be pushed off center stage at the time of the rise of the great patriarchal societies. The dominance of the male in society was accompanied by the assumption of central power by the sun gods or sky gods in myth and worship. Egyptian divine kingship, Sumerian militarism, Hebrew monotheism, and Greek rationalism suppressed worship of the Moon goddess.

In Egypt the institution of divine kingship gave rise to monotheism, the worship of the Sun god Ra, whose earthly incarnation was the Pharaoh of Egypt. This was the first cultural institution to express the *total* dominance of the patriarchal principle. Rationalism and militarism fought the influence of the moon goddess, in her diverse forms, throughout the ages. So strong was her appeal, however, that vestiges remain.

The appeal of the Moon goddess is easy to understand. There are forces in our emotional makeup, in our unconscious minds, that are strong, mysterious, and, at one and the same time, frightening and beautiful. These are the wellsprings of creativity, love, and fertility. The worship of the Moon goddess centered around these inspirations and gave socially accepted form to our emotional, irrational side. Today, people who have difficulty coming to terms with their own emotional disarray often pay high fees to a psychiatrist. It is the price paid in a world where only rational behavior is socially accepted.

The need her worship fulfilled made the Moon goddess a universal and integral part of *our psychic history*. We owe to her the recognition of a basic paradox: the irrational is a necessary component of a whole and balanced life. Without the intuitive and even the irrational, things would remain the same;

there would be no real change. The Universe would proceed soberly like a great complex and inanimate computer. The mechanistic view of the Universe is the ultimate expression of rationalist thought.

The symbolic importance of the Moon in psychic history, its dynamic presence in the evolution of the species, and recent discovery of its physical power in our lives point to the necessity of expanding our consciousness of the Moon—*our lunar knowledge*. Of immediate importance is the role of the Moon, physical and symbolic, in our emotional lives.

The persistence of lunar knowledge and tradition in the face of scientific and historical prejudice is heartening. In the 1930s, as scientific rationalism was busy taking over the world in the name of progress and attempting to finish off any remnants of lunar knowledge, M. Esther Harding wrote *Woman's Mysteries* in protest against monolithic rationalism. The book was published shortly before the dominant nations blundered into the Second World War, and recently was reissued. I hope people will pay more attention to it this time. Ms. Harding stresses that without the feminine principles of inspiration, intuition, and creativity our society loses balance and emotional value— resulting in an inability to avert war and disaster. The Moon's inspiration shows by comparison the emptiness of pure logic: "Ideas formed under the Moon," she wrote, "inferior though they may seem to be, have a power and compelling quality that ideas originating in the head rarely have."

The Moon has symbolized the poet's inspiration for as long as human memory extends. The power that lies in the sublime beauty of lunar inspiration, however, can become very dangerous if repressed. Turned inward upon the self, an individual's creative drive becomes distorted and grotesque. The repressed creative side of rigid rationalists is their own undoing, causing them to pursue, with logical doggedness, goals like absolute power. Ms. Harding's concern with the balance of male and female principles—Sun and Moon—corresponds with some findings of Jung.

In trying to restore the psychic balance of his patients, Jung found that both masculine and feminine principles were basic components of the structure of the psyche. An individual's conscious sexual identity would be balanced by the opposite principle in the unconscious mind. In a healthy individual this results in psychic equilibrium. We must treat our unconscious minds and their talents with respect if we wish to keep our head together. What both Harding and Jung imply, of course, is that society as a whole is out of balance and courting disaster.

The historical, patriarchal suppression of the lunar mode of life has eroded our social and individual wholeness. Tremendous material gains were made by Western science. But the planet has been raped and plundered by modern production methods and the balance of nature totally disregarded. With logical thoroughness, man has worked his way into a dangerous situation. *The current desperation of our society can be seen most clearly by moonlight.*

It is not accurate to declare that the Moon *causes* madness and crime. But it is precisely accurate to say that it is the repression of the Moon's influence and what the Moon stands for that brings about social tension, disharmony, and lamentable, often bizarre, results. It is a killer Moon for individuals who are *not* psychically balanced or for a society too rigid to roll with the cosmic punch.

Our ancestors applied their lunar knowledge in an attempt to live in a less frustrating relationship with the Universe. We must, in our own way, do the same. We are only at the beginning of a modern attempt at defining and applying lunar knowledge. The need for such development is widely felt, consciously and unconsciously.

The extent of the need can be judged by the overwhelming response of the general public to documentation of our research findings on lunar influence in human aggression.* A popular belief, held by members of the scientific community and

* Prior to the formal publication of my paper in 1972, an article appeared in the Miami *Herald* that discussed the results of my research (after I had made a presentation at an international scientific meeting). The Associated Press picked up on the news story, transmitting it to newspapers around the world. There was an immediate public reaction. The news media, particularly radio and television, were eager to broadcast the findings.

the public in general, seems to have been validated. It resulted in general enthusiasm for the work. The need for lunar studies seems to be recognized across all strata of society. The conscious response to the exploration of lunar knowledge is paralleled by evidence we see around us of the unconscious expression of this need. Millions of people consult astrology books and horoscopes in newspapers and magazines to fill a gap in their view of themselves. Even though most people realize that horoscopes will not prove accurate, consulting an "oracle" makes them feel more secure. Society's failure to come to terms with the inituitive side of human nature has resulted in a vast market for cosmic pseudoknowledge.

9
The Biological Tides Theory

In recent years many researchers from diverse scientific disciplines explored the relationship between the Sun and life on Earth. The Sun produces the energy that makes our planet hospitable to life as we know it. Energy from the Sun is transmitted Earthward by way of radiation ranging across the entire electromagnetic spectrum. The protective outer atmosphere of the Earth screens and absorbs many radiations that would be dangerous to life. An understanding of the effects of electromagnetic forces on all terrestrial phenomena, including on life itself, has been essential in the understanding of man's place in the Universe. As indicated in earlier chapters, electromagnetic forces have been intimately associated with the evolution of life, the development of the various species of life, and the generation of biological rhythms of metabolism that are essential for the survival of all terrestrial organisms.

The Moon does not produce radiation; its light is only reflected sunlight. It circles the Earth in a regular monthly course. The gravitational attraction between Moon and Earth varies according to the relative positions of Moon, Sun, and Earth. The Moon's pull is the strongest when the Moon, Sun, and Earth are roughly aligned, as in the phases of new and full moon, and its pull is weakest when the three bodies form a right angle, as during the quarters of the moon.

The daily and monthly lunar gravitational cycles, which produce our ocean and atmospheric tides, play an essential role in life processes. The Moon also produces a regular variation in the Earth's magnetic field generated by the daily lunar transit. The amplitude of this daily effect changes in a monthly cycle according to the phases of the Moon.

Solar researchers, in measuring electromagnetic field changes relative to the sunspot cycle, have invariably noted lunar rhythms in their data. Obviously, the Sun and Moon interact in some way. The net effect of the interactions has been measured in atmospheric phenomena, weather variables, terrestrial geomagnetism, life processes, and animal and human behavior. In an effort to understand the role of gravity in life processes and to determine how gravity interacts with other forces of the Universe in regulating evolution, growth, development, and behavior, I developed my theory of biological tides.

The theory of biological tides is based on empirical observation, five years' research, and a synthesis of findings in physics, astronomy, meteorology, biology, environmental ecology, and psychology. It attempts to place in perspective diverse observations and speculations. Such a theory must be able to account for facts already observed and documented. It must survive further testing, and it must predict future observations. Even if a theory, once constructed, proves wrong when tested, it has nevertheless been instructive. If we wish to learn more about the world in which we live, assumptions are justified in the face of uncertainty. We must be prepared to venture into the unknown.

Before explaining my biological tides theory, some background information may be necessary. The nine planets in our solar system revolve in elliptical orbits around the Sun. Our Moon travels with the Earth in its annual orbital odyssey. The Sun is one of millions of stars in the Milky Way Galaxy. Its nearest neighbor star, Alpha Centauri, is about 4.3 light-years away. The solar system revolves in an elliptical orbit about the center of the Milky Way's axis. The Milky Way is one of countless galaxies, all of which are believed to be moving in a radial direction away from one another and outward from what is presumed to be the center of the Universe. This outward radial movement of galaxies is thought to be the result of the Big Bang. The Big Bang theory states that the Universe evolved, about fifteen billion years ago, from the primordial explosion of a dense core of matter. At some point in time, the galaxies (it is thought) will stop moving outward and begin moving back inward, in a massive implosion, eventually bringing about another Big Bang. (Giant radio telescopes have detected signals from beyond our galaxy that are thought to represent the residual sound of the primordial explosion.)

Within this cosmic context we will concern ourselves primarily with our own solar system. All movements in this system occur in fairly predictable cycles. The Sun rotates on its axis once every twenty-seven days. Matter is converted to energy in a regular cycle of eleven years, measured by the number of sunspots observable through optical telescopes. Sunspots are centers of intense magnetic activity. When a group of sunspots coalesce, an unbelievably hot explosive region of activity known as a solar flare may develop. These flares emit vast amounts of X-rays, ultraviolet radiation, visible light, and long-wave radio energy. Highly charged particles from the Sun reach the Earth in a little over eight minutes. Longer wavelength radiation may take twenty hours to several days before reaching Earth. The path of these particles is deviated by the Sun's magnetic field, which extends downward to intersect with the magnetic field of the Earth.

The more energetic radiations penetrate the Earth's magnetic field and reach the surface, often playing havoc with electrical equipment; for example, they've caused circuit breakers for whole towns and cities to go out. Ultraviolet rays and X-rays disrupt the ionized layers of the atmosphere and interfere with broadcast signals. The solar wind is a continuous outpouring into space of ionized particles of gas from the Sun's corona. These particles enter the Earth's atmosphere chiefly in the polar regions and are related to the aurora borealis, "the northern and southern lights." The solar wind is gusty and

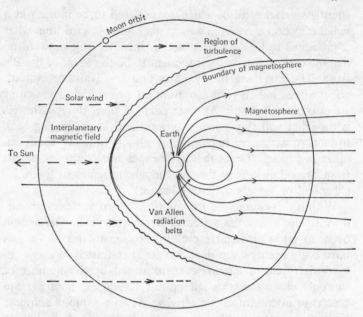

Magnetic field of Sun intersecting with magnetic field of Earth.

gives rise to periodic distortions of the Earth's magnetic field.

During the time of solar-flare activity, when the Sun is especially turbulent, solar wind blows rapidly and energetically through space. During solar storms, there is a dramatic increase in world-wide metabolic activity among humans and animals. At the time of a quiet sun, on the other hand, the solar wind settles down to a "steady breeze."

The Earth, which rotates on its axis once every twenty-four hours relative to the Sun (solar day) and once every 24.8 hours relative to the position of the Moon (lunar day), continually revolves in this fluctuating electromagnetic environment. The Moon, orbiting every 29.5 days (the synodic month), continually intersects the Earthward flow of electromagnetic radiations from the Sun. The Moon causes daily and monthly variations in the Earth's magnetic field; the variations can be considered to be *magnetic tides*.

The main forces of the Universe are nuclear forces, electromagnetic forces, and gravitational forces. Nuclear forces are strong, but operate over minute distances. The electromagnetic forces have been mentioned previously. We live in an electromagnetic world. Everyday materials, including the tissues of the human body, are held together by electromagnetic forces that operate between atoms. Human and animal organisms, viewed in a simplified fashion, can be thought of as a welter of electromagnetic forces separated only by their skin from the electromagnetic forces of the environment. The skin is a semipermeable membrane that permits movements of electromagnetic forces *in both directions*, maintaining a dynamic equilibrium between the internal and external electromagnetic fields. Gravity is, in one respect, the strongest of the universal forces. It ties the Earth to the Sun and the Moon to the Earth. The force of gravitational attraction between two objects depends upon the mass density of the objects and the distance between them. The greater the density, the greater the gravitational attraction; the greater the distance between two objects, the less the gravitational attraction. When the gravitational force is strong, the nuclear force is weak, and vice versa. Nuclear forces are ineffectual over long distances. Gravitation forces, though modest at short ranges, control the spatial relationship and movements of every body in the Universe. Gravity brings the matter of a star together and keeps it hot. Gravity helps generate the heat necessary for the formation of the very elements of which we and our planet are composed.

All life and all astrophysical phenomena are manifestations of nuclear, electromagnetic, and gravitational forces. Sometimes the forces collaborate harmoniously. Sometimes they collide unevenly and produce catastrophic results.

At the time of his death, Albert Einstein was concerned with the establishment of a unified field theory. He tried to comprehend the known forces of the physical universe as continually interacting parts of one large unitary field. That is, the known forces would be thought of as inseparable, acting together always. The concept of a unified field theory helps in thinking about the interaction of solar and lunar cycles and facilitates the understanding of biological tides. From the viewpoint of unified field theory, the interconnection of the following events becomes plausible, if not inevitable:

—The atmospheric tides in such phenomena as precipitation totals, variations in barometric pressure, tropical storm and hurricane formation;

—Lunar periodicity in the ion and electrical currents of the ionosphere;

—Lunar rhythms in the fluctuating magnetic field of the Earth;

—The relationship between human behavior and ambient ion currents during periods of hot dry winds.

The Universe may be viewed as an open system encompassing a hierarchy, with the Universe at the top and the biosphere

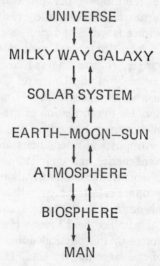

THE UNIVERSE AS AN OPEN SYSTEM

UNIVERSE
↓ ↑
MILKY WAY GALAXY
↓ ↑
SOLAR SYSTEM
↓ ↑
EARTH—MOON—SUN
↓ ↑
ATMOSPHERE
↓ ↑
BIOSPHERE
↓ ↑
MAN

Events occurring at any level of the hierarchy
have force-field reverberations extending to all
the other levels.

and life at the "bottom." Every element and force of the open system interacts with each and all of the other elements and forces. Events that occur in any one part of the system have reverberations in all other parts. Not only do events that occur in the biosphere affect life on Earth, they also radiate outward through the system into other areas. Solar flares cause magnetic storms that impact upon the biosphere, affecting life processes. Radio transmissions, nuclear experiments, man-made electrical currents, and even the use of aerosols cause changes in the immediate biosphere, in the upper layers of the atmosphere, and perhaps in the Sun-Moon-Earth system itself. We are an integral part of the constantly changing Universe, and must ebb and flow with its fluctuations in order to survive.

In an earlier chapter, I presented the analogy of the human organism being composed of the same elements, and in similar proportion, as the surface of the Earth. It is reasonable to speculate that the human organism behaves in the same fashion as does the Earth in response to gravitational forces and the geo-electromagnetic fluctuations. My biological tides hypothesis, in its simplest form, states that the human body is susceptible to the same cosmic influences as the Earth and that body processes ebb and flow with the tides, just as do the crust and the waters and the magnetic field of the Earth.

Gravity exerts a direct effect on the body and its water mass and an indirect effect which is mediated by the electromagnetic field. Both effects are believed to act at two main sites—the body water and the nervous system. Because the body is composed of 80 per cent water and 20 per cent "land," or solids, it is reasonable to assume that gravity exerts a direct effect on the water mass of the body, just as it does on the water mass of the planet. A biological tide would take place on predictable daily and monthly bases. Times of excessive gravitational force (i.e., coincidence of several cosmic cycles) might yield excessive body tides. These tides could be expected to dissipate quickly in a well-functioning system and result in prompt relief of fluid tension.

The indirect effect of gravity, mediated by the Earth's electromagnetic field, occurs simultaneously with the direct effect. Environmental noise resulting from any of a number of extraterrestrial events (for example, sunspots, solar flares, solar wind, cosmic rays, or excessive gravitational tides due to special planetary configurations) cause severe disturbances in the ambient electromagnetic field that bathes all human organisms. Electromagnetic-field disturbance results in a Piccardi effect on the body water. The physical properties of the water and the distribution of the colloids, which are in suspension in the body fluids, undergo a sudden and dramatic change. This change in the physical nature of the water, together with altered cell membrane permeability (the Dubrov Effect), results in altered fluid-flow dynamics among the body's three water compartments. The very nature of the body water itself has been changed, and the fluid flow across cell membranes is thrown out of kilter. In such cases, bloating and discomfort could last for several days.

The second site of gravity's influence is presumed to be the body's nervous system. Here, too, I believe that a direct effect of gravity and also an indirect effect mediated by the Earth's electromagnetic field take place. Regarding the indirect effect, Dr. Robert Becker postulates that fluctuations and disturbances in terrestrial electromagnetic fields are picked up by amplifier nodes along the body's nerve circuits. These amplifier nodes then modulate nerve conduction. The electrical conducting properties of the nerve itself are altered and the threshold of neural firing is changed. These influences can make the nervous system more irritable and susceptible to rapid firing or less irritable and more sluggish than normal. The change in condition that comes about may well be dependent upon the state of the organism's receptivity—positive or negative.

Little is known about the direct action of gravity on the body's solid mass. It is likely that this effect is also mediated via the nervous pathways. It is my speculation that we shall discover there are "gravoreceptors" within the human body. They might be located along nervous pathways and in the walls of the blood vessels. Gravoreceptors, picking up the immediate thrust of gravitational force, could mediate sudden direct shifts of nervous and vascular functioning.

Having studied the evidence concerning the function of the

pineal gland in humans, it is my speculation that the pineal may serve a gravoreceptor function as well. The pineal is thought to be an electrochemical transducer capable of receiving electrochemical impulses both from the body and from the environment, converting them into pineal hormones—melotonin and serotonin. These hormones have an effect on biological rhythms of the sexual and the reproductive cycles. Is it possible that the pineal gland, receiving gravitational and electromagnetic information from external and internal environments, synthesizes this information and sends out signals to the blood vessels, nerves, and sexual organs? If so, these signals could serve to control biological rhythmic processes and to cushion the organism against sudden internal discord and disharmony resulting from the impact of intense environmental disturbance. In this sense, the pineal gland can be viewed as an environmental shock absorber.

We can trace the effects of biological tides using the general-systems theory and the unified field model. At one end of the outline we have the microscopic view of the body's internal environment in relation to the external environment. The internal environment consists essentially of bundles of cells bathed by extracellular water. They are energized by nerve impulses that are electrochemical processes coursing along a multitude of nerves—the body's electrical circuitry. As each nerve impulse proceeds along its nerve fiber, a small electromagnetic field is generated in a circular fashion surrounding the fiber. On reaching the end of the nerve fiber, the electrical impulse triggers the release of chemicals known as neurotransmitters. These chemicals carry the neural impulse either to another nerve fiber or to a target cell, which may be muscle, gland, blood vessel, and so on. These neurotransmitters (dopamine, acetylcholine, serotonin, norepinephrine, gamma aminobutyric acid) act on the target cell in such a way as to excite or inhibit its functions.

Let us examine the other end of our cosmobiologic system. In a rhythmic fashion, cosmic forces constantly play upon the Earth. Disturbances from the Sun or from other galactic sources cause a massive distortion in these rhythmic functions. This distortion is known as "noise." The disturbances impact upon the ionosphere, which forms the outer protective layers of the atmosphere surrounding the Earth. Changes in the ion and electromagnetic balance of the ionosphere caused by cosmic impacts are transmitted through the atmosphere to lower levels. Eventually, they disturb the electromagnetic equilibrium of the atmosphere at ground level.

Noise can arise from any of several sources. There can be electromagnetic noise and gravitational noise; the one interacts with and reverberates with the other. According to unified field theory, all of the forces resonate and interact with each other.

Let us say that there is a coincidence of cosmic cycles—for example, a full moon coinciding with lunar perigee and a lunar eclipse. Sun, Moon, and Earth are aligned in a geometric plane. At such a time, the force of gravity bearing upon the Earth is increased well over the usual daily gravitational tide. In addition, the magnetic field of the Earth will be greatly distorted by the magnetic fields of the Sun and the Moon with which the planet is aligned. Resultant electromagnetic noise impacts on our ionosphere and plays havoc with ion and electromagnetic currents. These forces impinge upon the biosphere. The human organism is bombarded suddenly by a massive disturbance of gravity and the surrounding electromagnetic field. The disturbance dramatically shifts the equilibrium between our inner and outer worlds.

The sudden shift stresses our internal systems. As described previously, the nervous system may become irritable, altering thresholds of nerve firing. There are buildups or deficiencies of water in different parts of the body.

People who are healthy and emotionally well balanced tolerate added stress with a minimum of physical or emotional discomfort. Even when there is an immense disturbance caused by extremely excessive noise, a new equilibrium is rapidly achieved. Any discomfort is fleeting and mild. Such environmental noise may have made you feel, from time to time, that it was no use trying to get work done or that it was hopeless trying to communicate with someone. This type of passing frustration is one example of what happens as a result of a shift in cosmic forces. As a result of increased cosmic stress, many people experience mild headaches, transient depressions, grumpiness, and insomnia. The feeling that "this just ain't my day" may sometimes be extraterrestrial in origin.

People with unstable personalities and mood disorders or those who are already under considerable physical or emotional stress may experience major setbacks during periods of excessive cosmic stress. If they are violence-prone, they may be triggered into uncontrollable behavior.

Many aspects of the biological tides theory have been documented by the work of other researchers. F. G. Sulman, working in Israel, has demonstrated that the rise in physical and mental illness at the times of the sharav, which occurs there twice each year, is a result of the excess positive ions at earth-level. The excess causes increased production of serotonin in the central nervous system. A resulting rise in blood pressure due to constriction of the smaller blood vessels would influence the exchange of fluids among the fluid compartments of the body. Serotonin is thus linked to the workings of the biological high-tide situation in the body by influencing the pressure in one of the fluid compartments.

There is another fascinating serotonin connection. The biologist Harry Rounds analyzed the lunar rhythm of heart-accelerating substances in the blood of roaches, mice, and men. He suspects that one of the heart-accelerating substances whose presence in the blood follows a lunar rhythm is serotonin. If this proves true, the presence of serotonin in significant amounts in the blood immediately after new and full moon could explain a good deal of the fluid imbalance and nervous irritability associated with the biological high-tide syndrome. It would not be surprising if research on the other neurotransmitters ultimately demonstrates a lunar periodicity in them as well.

The finding of excessive bleeding tendencies in humans at the time of new and full moon, as reported independently by Andrews, Rhyne, and others, provides further supporting evidence for biological tides.

I believe that someday we shall be able to tie in the effects of electromagnetic forces on *specific* body processes. Dr. Dubrov has explored the influence of electromagnetic fields on cell-membrane permeability—another probable mechanism at work in the biological high-tide syndrome. Dr. Becker believes the nervous system is directly affected by shifts in the electromagnetic fields. His view is supported by the work of other researchers, notably Presman, Lang, Persinger, and Ossenkopp.

By now it should be clear that man has been continually interacting and evolving with the Universe and that the closer one looks at relationships, the more connections one can see. Our studies, as well as many others that have appeared in the literature in the last few years, support a concept of man in dynamic equilibrium with the Universe. This seems a perfectly reasonable view of the nature of man in relationship to the physical universe. It is a unitary concept totally compatible with the views of major scientists and philosophers through the centuries—particularly with Einstein's unified field theory, with Darwin's theory of evolution, and with Ludwig von Bertalanffy's general-systems theory.

From a unified interaction perspective, I believe that a new field of science is indeed evolving before our very eyes—cosmobiology. It is a unifying border science that will clarify the relationship of man to the Universe according to the laws of nature as we know them or as we are finding out about them.

A number of terms are applied to researchers and workers in the various fields that attempt to link the biological reactions of man and animals to changes in the environment. A partial listing of these terms includes medical climatology, biometeorology, chronobiology, meteorpsychiatry, ecobiology, and biomagnetism. Each of these fields is in itself somewhat narrow and confining, looking at only certain aspects of the overall picture. Because so much of the research in these fields has been accomplished only in the last two decades, and because the research has been pursued and reported in many different countries and in a variety of languages, a synthesis allowing the emergence of an overall perspective had not been achieved previously. What I have attempted to do here may represent the first synthesis of an overall perspective of cosmobiology.

The term cosmobiology was used by Dr. W. F. Petersen in the 1940s; it may have been coined by researchers working much earlier in the century. In my opinion the term *cosmobiology* denotes nicely the interdisciplinary research efforts of those who are interested in the relationship of man to the natural geophysical environment and to the cosmos in general.

A theory, in order to be credible, must account for all of the relevant observed phenomena. We have noted that geographical latitude and longitude play a role in the timing of biological cycles. This was evident, for example, in the lagged homicide peaks in the Cleveland murder sample. Research into the cycles of the growth and movement of animal populations disclosed latitudinal lags. Leonard Wing, an early researcher of rhythms, termed this phenomenon "latitudinal passage" and described it as operating in a wide range of naturally occurring cycles. Other researchers noted phase-shifting in the biological cycles, especially of birds flying from northern latitudes to the equator as winter approaches and in the reverse direction during summer.

Precise mapping of the effect of geographical location on the timing of lunar rhythms of behavior would require extensive research studies. For any given behavior, the ideal situation would be to study the behavior in several cities at the same geographical latitude but at different longitudes and, also, in several cities at the same longitude but different latitudes. The vagaries of official data-collection from place to place virtually preclude such efforts. Another approach would be to conduct animal studies at different locations. This would require an effort on an international scale. Hopefully, experiments at different locations will go forward as a test of the biological-tides theory.

Einstein viewed time and space as being one and the same continuum. To us, space has three dimensions—latitude, longitude, and height above the center of the Earth. Time is also multidimensional. For our purposes, we may consider solar time, lunar time, and sidereal time as ever-present and continually interacting.

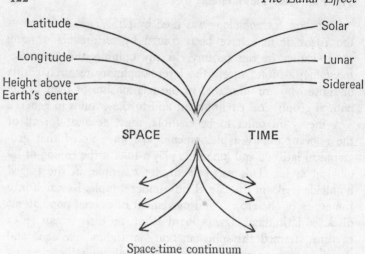

Space-time continuum

The electromagnetic field at earth level is constantly fluctuating and varies with space and time. There are no constant
conditions in terrestrial magnetism or in nature in general. This
explains why studies performed in different geographical locations yield different results—and why studies done at the same
location but at different times yield different results.

Biological time is relative—a reflection of the net interaction
of all environmental variables impinging upon the organism at
its unique moment in space and time.

The most convincing evidence for the existence of a lunar
component in human biological rhythms came to my attention
as *The Lunar Effect* was going to press. At the end of 1977, Dr.
L. E. M. Miles and associates of Stanford University School of
Medicine reported their investigation of the biological rhythms
of a psychologically normal twenty-eight-year-old man, blind
since birth. This man, who lived and worked in a sighted society, suffered cyclically from a severe sleep-wake disorder. For
two to three weeks at a time, over several years, insomnia and
excessive daytime sleepiness would interfere with his work and
leisure activities.

Prolonged monitoring in a sleep laboratory of the patient's brain-wave activity and a time-series evaluation of his hormones and body chemistry revealed a dramatic fact. When he was isolated from all solar- or calendar-time cues, the patient demonstrated a lunar-day rhythm (24.9 hours) in all measured biological functions. His circadian rhythms of body temperature, alertness, performance, cortisol secretion, and urinary electrolyte excretion were desynchronized from the 24-hour societal schedule. They showed a persistent period of about 24.85 hours. The sleep pattern became normalized, and synchronized to a 24.9-hour period. Remarkably, there was a coincidence between onset of sleep and a local low tide.

The fact that a person who is isolated from *all* solar-time cues assumes a lunar mode of biological functioning is further proof that human circadian rhythms are composites reflecting the net effect of both solar-timed and lunar-timed rhythms—which is confirmation of the concept of biological tides.

10

Applied Lunar Knowledge

How should society apply what is known about the power of the Moon?

My experience is that people who are stable and well balanced may be minimally affected by cosmic influences such as moon phases or sunspots. To be aware constantly and always on the alert for cosmic influences might have an adverse effect. It certainly introduces a distraction we don't need. But are there certain people who *should* take the Moon into account regularly? There are a large number of people living in every city and town who are tiptoeing on the edge of reality. They may be alcoholics, drug addicts, violence-prone or accident-prone individuals, or criminally-inclined types. These persons often have wide mood swings, poor impulse control, and difficulty in controlling their emotions. They tend to live in a continually unstable emotional state. For these people, any stress over and above what they usually have to cope with on a day-to-day basis could be the last straw.

When the pattern of imperceptible environmental forces is disturbed suddenly, dramatic behavioral and emotional changes can result. Because of the possibility, it would be wise for doctors, sociologists, police, fire-rescue, and hospital personnel to be alert to the triggering potential of cosmic stresses. We should bear in mind that a wide variety of stresses in the natural environment—gravitational, electromagnetic, atmospheric, and so on—can act as triggers.

Poorly balanced people are affected in a way no different from you or me. *None of us can escape the pervasive ambient geophysical environment.* We are constantly washed in its electric, magnetic, and ionic fields. We are bombarded constantly by radiation. Most of us have a reasonable stress tolerance and can cope with the minimal increase in stress resulting from external environmental changes. One can better understand the plight of the individual with poor impulse control and low stress tolerance by thinking of him as standing on a tightrope. If a person is standing on the ground with his feet well planted, you could shove him and chances are he won't tumble over. But if we put the same person on a tightrope, he will have great difficulty balancing, not being experienced at tightrope walking. A shout or a scream or even a breeze might be enough to make him fall. It doesn't take much of a disturbance at all, as you can see, if balance is precarious.

People arrive at a state of poor mental and emotional balance via a wide range of routes. Some may have precarious personalities by virtue of intensely traumatic early childhood and family experiences. They may have no basic biological or chemical disorder, but are conditioned in such a way that they do not have well-integrated personalities. On the other hand, it is known there are genetically determined biochemical differences in certain people that can predispose them to poor emotional balance. It is known, for instance, that schizophrenia and manic-depressive illness are hereditary disorders. Manic depressives have hereditary susceptibility to widely fluctuating moods. Some of them behave in an extremely bizarre manner. Exactly when a manic depressive is likely to "go off the deep end" has been a matter of conjecture. Recent research indicates some of them are more inclined to become maniacal at times of new and full moon.

Schizophrenics and borderline personalities—those who tend to become psychotic under stress—are known to have a hereditary predisposition to their disorders. Psychiatric researchers are beginning to investigate these genetic defects, to pinpoint them, and to discover the actual biochemical abnormalities present in the nervous system. What pushes these persons over? Just about any stressful event. It could be a family argument, separation from a loved one, or the loss of a job. It could be any kind of stress, including cosmic stress.

We have investigated just one type of stress. In practical terms, cosmic stresses may present an advantage to us when we deal with them because they are predictable. Conceivably, we can try to devise ways to forecast and to counter the effects of cosmic cycles.

We have chemicals and medications that are capable of preventing stress-overloading. There are a variety of tranquilizers that can alter the threshold of irritability and, therefore, change the response to stress. There are strong diuretics that can be used to treat excessive buildup of body fluids, reducing tissue tension and restoring a more or less normal fluid balance in the body. The use of the mood-stabilizing drug lithium carbonate in the treatment of manic-depressive illness has brought about dramatic improvements in the lives of persons having this mental disorder. Lithium carbonate is thought to act on the postsynaptic nerve-cell membranes. If there is excessive nerve firing because of an episode of mania, lithium carbonate will modify the firing mechanism and stabilize mood and body water. We have the tools and methods to treat biochemical imbalances that set people up for aberrant behavior.

I see in my office a number of manic depressives who are stabilized on lithium. They usually do well and are symptom-free. On occasion, many of them call in *at the same time,* all reporting recurrence of similar symptoms—restlessness, depression, insomnia, rapid heartbeat, feeling "off the wall." I order blood tests, and invariably blood lithium levels are below the therapeutic range. A temporary dosage increase alleviates the symptoms. A few days later they can be returned to their previous daily dosage without fear of symptoms recurring immediately. I am certain the only reason for this pervasive phenomenon is that these patients respond en masse to an atmospheric perturbation. Most likely it is lunar-related, but a sudden increase in sunspot activity or a cosmic ray shower can produce sudden metabolic imbalance.

Patients readily accept my explanation. They seem reassured that it is not the lithium that is failing them. In my experience there is no stable daily-maintenance dose of lithium. Treatment must be altered from time to time to counteract sudden imbalances in the geomagnetic environment to which manic depressives are exquisitely sensitive.

We are now confronted with a thorny philosophical question. How can we justify preventive application of these methods?

Examine the crime rate in any city! We have ghettoes where people are poor, overcrowded, and angry. Handguns are readily available. Violent behavior is a regular and periodic phenomenon. Should we treat everybody with drugs? Should we identify violence-prone individuals by their police records and force them to take pills against their will at times of known increases in gravitational and/or geomagnetic stress? This "solution" could generate more violence than it prevents. Such "solutions" smack of Hitlerism. They are especially reminiscent of Anthony Burgess' frightening story of totalitarian behavior control, *A Clockwork Orange*.

There are many people who believe that forced and involuntary behavior control should be used on anyone who is identified as an "antisocial personality." There are also many people who believe that we have no right to control the behavior of any human being. The truth is, society does *not* have the legal right to control behavior against the will of an individual —and everyone ought to be aware of this fact!

Schemes of drugging so-called antisocial individuals are often motivated more by a desire for revenge than by good sense. The fact is, the great majority of people who commit murder are *not* known habitual criminals! Most of them don't have a criminal record. They had never before committed a crime. Habitual, recidivist criminals represent a small element among those who commit murder. By far the most frequent source of criminal violence is domestic conflict occurring between family members or friends, especially when alcohol is involved. We can't tell in advance who is going to be murderous tonight.

Happily, there are ways in which our knowledge of cosmic stress and its effects can be sensibly used. When a psychiatrist's or a psychologist's patient shows periodic outbursts of maniacal behavior but is otherwise relatively normal, we may be able to apply what we have learned about the triggering of such behavior. Such a pattern of behavior was originally thought to be moon madness. It was called *lunacy*. The word lunacy is derived from the Latin *lunaticus*. This term was frequently used by Roman physicians and philosophers. It was most likely first applied to epileptics who had been observed to have more seizures around times of new and full moon. Lunacy came to be applied in general to people who exhibited periods of maniacal behavior interspersed with long, lucid periods of normal behavior.

Correlating with the ancient concept of lunacy is the well-known syndrome of circular manic-depressive illness. This is characterized by periodic episodes of wild, bizarre, hyperexcitable, and irrational behavior. Between episodes, most manic depressives function in a normal, productive fashion. They show no evidence of thinking disorder, of psychosis, or of a major mental illness of any kind.

Studies, notably the one published by Michael H. Stone, M.D., in *Psychiatric Annals* in 1976, have shown that the incidence of manic episodes tends to increase in relation to new and full moon. The development of an effective treatment for manic-depressive illness using lithium has allowed for greater insight into the neurochemical mechanism of this disorder. Lithium, as noted earlier, stabilizes the firing of nerve cells, balance of body water, and mood. Its mechanism of action is compatible with the suggestion that episodes of mania correspond to an extreme biological high-tide situation in the body. Correction by lithium suggests that a dangerous biological high tide is treatable by addition of a simple chemical to the body fluid.

This is very exciting. It is entirely possible that the episodic dramatic behavior of precariously balanced people may be due as well to extreme biological high tides. At present, the experi-

mental use of lithium extends to preventive use for recurrent
depressions, cyclical mood swings, and episodic aggressive behavior. It may provide the answer to the question of what
can be done about altering the pathological effects of solar-
and lunar-related physiological, biochemical, and behavioral
changes.

For many patients, we may be able to avoid constant tran-
quilizer doses by merely restoring normal fluid balances, making the person less susceptible to subtle triggering forces both
in his social life and in the natural geophysical environment.
On the doctor-patient level, this knowledge should now prove
extremely useful.

We may be able to use our knowledge of cosmic stress in a
constructive way on the broader social scale as well. If we can
predict a period of general behavioral upheaval in a given area,
we should be able to design useful anticipatory programs. In
fact, my pilot prediction study, using a coincidence of cosmic
cycles as its cue, predicted a time of general upheaval. Reports
of mortality, accidents, and mental illness rose significantly.

The measures that can be taken are simple and sensible. The
public could be informed by the news media to expect an increase in irritability and distraction that can lead to accidents
and arguments. Police, fire, and ambulance units can be alerted
so that their resources are mobilized during danger periods. Hospital emergency rooms could be alerted so that they are ade-
quately staffed.

The objection might be raised that issuing such alerts and
warnings could generate a degree of panic. In my opinion this
is unlikely. As I have emphasized, police, fire-rescue, and emergency-room personnel are for the most part already aware of
the influence of the Moon on human illness, accidents, and behavior. Confirming the effect would only validate their own experience. People would be more confident, especially if, for example, hospital authorities saw fit to augment the staffs of their
emergency rooms at such times.

It is important that we put cosmic triggering effects in perspective. Most violence is not connected with the Moon or

with other cosmic influences. Killer moon causes only a small part of the violence.

If you think of violence as a big American apple pie and you want to know what size is the slice attributable to the influence of the Moon, I would say it would be no more than 2 or 3 per cent. But when one considers the total population of the world, this amounts to a *significant* absolute number of violent acts. If something we have learned can help to reduce that level of violence, we have made a worthwhile advance.

What we have done in our research in Miami is to discover an independent variable influencing violent behavior. This should lead us, first, to be forewarned and, second, to explore—through the biological tides theory—treatments that redress the biotide balance. Medications can be timed to the periodic stresses that trigger aberrant behavior. Such relatively mild measures would be preferable to electric-shock therapy or the psychosurgery now being used on some patients who show periodic violent behavior.

In applying lunar knowledge, we should consider using a calendar showing both lunar and solar time. We would be able to recognize lunar and solar periodicities, especially their coincidences. Ancient civilizations used lunar calendars; the Israelis, the Chinese, the Hindus, and the Moslems still do. Interestingly, there is much less violent crime in these societies.

An easy-to-use lunar-solar calendar could be generated by computer methodology. It would allow us to know in advance on what days to anticipate periods of irritability, disturbed or erratic behavior, or proneness to accidents. If people of widely varied backgrounds and skills could easily determine periods of potential emotional disturbance, the usefulness of such a calendar becomes readily apparent. The calendar should be *both* lunar and solar because only the combination will give the full perspective of what is occurring on the cosmic scale.*

* The Lunar Society, famous in England in the 1770s, held monthly meetings near the time of full moon so the members might ride home by moonlight. The "lunatics" included Erasmus Darwin, James Watt, John Baskerville, and Joseph Priestley. They were fascinated by science and invention.

It is essential to remember that lunar periodicity does not refer only to events that occur around new and full moon. My research revealed that not all violent human behavior peaked at new or full moon. In fact, only homicides and aggravated assaults peaked at such times. The peak incidence of psychiatric emergency-room visits was at first quarter, with significantly decreased frequency at new and full moon.

Lunar-timed peak incidence of certain behaviors can be expected to vary with geographical location as well. In Cleveland the peak incidence of homicide occurs approximately three days *later* than new or full moon. This reflects the lag in the tidal cycle that occurs in that area. Our data cannot be generalized to any area without a survey first being conducted in the area. It is reasonable to expect that different peak incidences would be encountered in each location.

The role of geographical location is not yet clearly understood. A number of studies spread widely over the globe would help to clarify the effect of latitude and longitude on lunar timing. We might find confirmation of Professor Wing's hypothesis of latitudinal passage in relation to cyclic occurrences in animal populations. In fact, prospective studies of animal behavior, such as those reported by F. A. Brown, M. Klinowska and D. S. P. Bisbee, are already underway in a wide variety of locations and are proving useful. Knowledge of the effect of latitude and longitude on these cycles would make it possible for professionals in different areas to make accurate reference to the interface of cycle and psyche.

In my psychiatric practice I often will ask myself what phase the Moon is in. In my work with geriatric patients especially, I have noted times when these people just don't do well. I have been able to correlate these times with one or another atmospheric disturbance, be it sunspots, coincidence of cosmic cycles, or a new or full moon. I am able to reassure my patients that their symptoms will be transient, that they are related to weather or atmospheric disturbances, and that they will soon subside. I have not had the occasion to medicate patients because of these symptoms. I don't rule out the possibility for the future.

I have found that older age groups in general are more susceptible to perturbations in the geophysical environment than are younger people. Older people are naturally more feeble. They are frequently ill and may have had one or more strokes. (We should note again that strokes have been linked to atmospheric disturbances.) Their general condition is weakened and they are more susceptible to stress of any kind, including subtle cosmic stresses.

I have had patients come to my office and announce, "I am a Moon person." They claim to be depressed or excited at times of new or full moon. Patients also tell me that they can't sleep during full moon. Some get out of bed on nights of full moon and wander around the house. I see people who claim they can tell *just by the way they feel* whether the Moon is new or full.

Some people who feel they are influenced by the Moon are also aware of their own violence-proneness. I have letters from people who are in prison for violent crime. They are not professional criminals. They are interested in understanding what might make them lose control in a given situation. If there is a way that they can learn to stabilize their moods and protect themselves from violent behavior, they would like to know about it. Our findings are potentially of use to these people.

Irrational violence, of course, has an almost infinite number of causes. We have discovered one significant variable that is, within limits, predictable. Everyone should know about it.

It would be useful for the general public to be educated as to our findings. People alert to high-risk times should beware of heavy drinking. They should lock up any weapons they possess. They should be extremely careful while driving. When there is to be a major coincidence of cosmic cycles and the natives are restless, my family and I take a simple precaution. Book-reading at home is a relatively safe pastime.

In view of the exposure my findings have received, I am sure there will be people dragged off to jail, after committing some heinous crime, shouting, "It was the Moon's fault, not mine." There are always people who will not take responsibility for their own behavior. They will blame it on anything and anyone they can. I have been contacted by a number of attorneys who were defending murderers or other violent criminals and seeking a diminished-responsibility defense for their clients. I wish to state here—unequivocally—that in my experiments I have found no evidence for a marked susceptibility of any given individual to cosmic influences. All my studies have been of a statistical nature. Murder statistics have been analyzed only as a group of data extending over many years. There is no scientific evidence to support a defense of diminished responsibility. In my opinion, this defense would be a misapplication and abuse of my research.

It should be obvious there is a wide range of possibilities for the application of lunar knowledge throughout society. The field of cosmobiology is new. We can expect many interesting developments in the future, which in turn will lead to further applications. At this time the most important applications of the material fall into two classes. First, there is application on the social scale. This involves alerting municipal services to cosmic-cycle coincidences so they can be ready to handle increased work loads. The second class involves application on an individual basis. When we foresee a time of cosmic stress or a coincidence of cycles, simple precautions should be taken so we are not in the wrong place with the wrong person(s) at the wrong time. Such a period would be inopportune for scheduling a day of driving in busy downtown traffic. A coincidence of heavy sunspot activity and a full moon might be the wrong time to ask the boss for a raise.

Mrs. Barbara Svens, director of health service at Franklin Pierce College, in Rindge, New Hampshire, has made a study of lunar timing. She suggests certain kinds of entertainment be scheduled for the quarters of the moon rather than during full or new moon. People, she found, are calmer during the quarters. A simple example of scheduling would be to book rock bands "and other excitement" for the quiet times of the month. During a full moon the city of Boston had an outdoor rock concert. The results were so disruptive that the city banned further concerts. Mrs. Svens properly asks, Would the concert have been so disruptive if it were held during the first quarter of the moon rather than during full moon?

Timing is the key word. With proper timing, and a bit of caution, we can learn to live in harmony with cosmic forces and with our natural environment.

To what extent do cosmic forces, like other determining factors in life, govern our fate? We are free within certain limitations to direct our lives. But it is not easily done. Man has chosen to view himself as different from lower animals because he thinks himself capable of self-determination.

In fact, it is the rare person who is able to determine his or her own direction in life. Self-determination requires enormous amounts of energy, drive, and motivation. Only one who is able to recognize, mobilize, and integrate basic drives is able to achieve relative freedom of will. The results of my research indicate that even those lucky enough to achieve some degree of autonomy are more subject to the vicissitudes of the natural geophysical environment than they would like to admit. People capable of recognizing and accepting the limitations of freedom are able to work within and around these limitations and thus exercise the best of human potential.

In other words, we are free only so long as we are able to recognize and accept the limitations of our freedom. This is the dilemma of free will trying to act in harmony with the Universe. How can we accept limits and yet, in other ways, transcend these limits? With pure reason alone we cannot hope to navigate through life in a smooth fashion. For sound judgments we must rely often on feelings and intuitions. Yet, paradoxically, the more we know of our limits, the freer we are because we will be less likely to waste time pursuing dead ends.

The Moon represents one of the many subtle cosmic forces that, by their configuration, determine the setting of our biological rhythms at birth and give us a pattern in which to synchronize our life with the Universe. It can also exert a pull on body and mind that cannot be denied.

From my own research and from a collection of other pertinent studies, I have assembled a great deal of evidence as well as speculation of a scientific nature on the undeniable influence of the Moon. The possibilities are fascinating, the leads promising. Nevertheless, scientific work in this field is in its infancy.

Occasionally it suffers from what Jung referred to as the "ruinous influence of statistics." We scientists must find languages and methods other than those of numbers and fleeting descriptions if we are to integrate a dynamic feeling for the Moon in our lives and in our work.

Ephemeris

PHASES OF THE MOON
(computed by U. S. Naval Observatory)

D = DAY
H = HOUR
M = MINUTE

NEW MOON	D H M	FIRST QUARTER	D H M	FULL MOON	D H M	LAST QUARTER	D H M
1978							
						Jan.	2 12 08
Jan.	9 04 01	Jan.	16 03 04	Jan.	24 07 56	Jan.	31 23 52
Feb.	7 14 56	Feb.	14 22 12	Feb.	23 01 27	Mar.	2 08 35
Mar.	9 02 37	Mar.	16 18 21	Mar.	24 16 21	Mar.	31 15 11
Apr.	7 15 16	Apr.	15 13 56	Apr.	23 04 11	Apr.	29 21 03
May	7 04 47	May	15 07 40	May	22 13 17	May	29 03 32
June	5 19 02	June	13 22 45	June	20 20 32	June	27 11 45
July	5 09 51	July	13 10 50	July	20 03 06	July	26 22 32
Aug.	4 01 01	Aug.	11 20 07	Aug.	18 10 15	Aug.	25 12 19
Sept.	2 16 10	Sept.	10 03 21	Sept.	16 19 02	Sept.	24 05 08
Oct.	2 06 42	Oct.	9 09 38	Oct.	16 06 10	Oct.	24 00 35
Oct.	31 20 07	Nov.	7 16 19	Nov.	14 20 01	Nov.	22 21 25
Nov.	30 08 20	Dec.	7 00 35	Dec.	14 12 32	Dec.	22 17 42
Dec.	29 19 37						

PHASES OF THE MOON

	NEW MOON			FIRST QUARTER			FULL MOON			LAST QUARTER		
	D	H	M		D	H	M		D	H	M		D	H	M
1979															
				Jan.	5	11	15	Jan.	13	07	09	Jan.	21	11	24
Jan.	28	06	20	Feb.	4	00	37	Feb.	12	02	40	Feb.	20	01	18
Feb.	26	16	46	Mar.	5	16	23	Mar.	13	21	15	Mar.	21	11	23
Mar.	28	03	00	Apr.	4	09	58	Apr.	12	13	16	Apr.	19	18	31
Apr.	26	13	16	May	4	04	26	May	12	02	01	May	18	23	57
May	26	00	01	June	2	22	39	June	10	11	56	June	17	05	02
June	24	11	59	July	2	15	24	July	9	20	01	July	16	11	00
July	24	01	42	Aug.	1	05	58	Aug.	8	03	22	Aug.	14	19	03
Aug.	22	17	11	Aug.	30	18	10	Sept.	6	11	00	Sept.	13	06	16
Sept.	21	09	48	Sept.	29	04	20	Oct.	5	19	37	Oct.	12	21	25
Oct.	21	02	24	Oct.	28	13	07	Nov.	4	05	48	Nov.	11	16	24
Nov.	19	18	04	Nov.	26	21	09	Dec.	3	18	09	Dec.	11	14	00
Dec.	19	08	24	Dec.	26	05	12								
1980															
								Jan.	2	09	04	Jan.	10	11	51
Jan.	17	21	21	Jan.	24	13	59	Feb.	1	02	22	Feb.	9	07	37
Feb.	16	08	51	Feb.	23	00	15	Mar.	1	21	00	Mar.	9	23	49
Mar.	16	18	57	Mar.	23	12	32	Mar.	31	15	15	Apr.	8	12	07
Apr.	15	03	47	Apr.	22	03	01	Apr.	30	07	35	May	7	20	51
May	14	12	01	May	21	19	17	May	29	21	29	June	6	02	54
June	12	20	39	June	20	12	32	June	28	09	03	July	5	07	29
July	12	06	46	July	20	05	52	July	27	18	55	Aug.	3	12	01
Aug.	10	19	10	Aug.	18	22	29	Aug.	26	03	43	Sept.	1	18	08
Sept.	9	10	01	Sept.	17	13	55	Sept.	24	12	09	Oct.	1	03	18
Oct.	9	02	51	Oct.	17	03	49	Oct.	23	20	53	Oct.	30	16	34
Nov.	7	20	43	Nov.	15	15	47	Nov.	22	06	40	Nov.	29	10	00
Dec.	7	14	36	Dec.	15	01	48	Dec.	21	18	09	Dec.	29	06	33

PHASES OF THE MOON

NEW MOON			FIRST QUARTER			FULL MOON			LAST QUARTER		
D	H	M	D	H	M	D	H	M	D	H	M

1981

Jan. 6	07	25	Jan. 13	10	10	Jan. 20	07	40	Jan. 28	04	20
Feb. 4	22	15	Feb. 11	17	50	Feb. 18	22	59	Feb. 27	01	16
Mar. 6	10	32	Mar. 13	01	52	Mar. 20	15	23	Mar. 28	19	35
Apr. 4	20	20	Apr. 11	11	12	Apr. 19	08	00	Apr. 27	10	15
May 4	04	20	May 10	22	23	May 19	00	04	May 26	21	01
June 2	11	33	June 9	11	34	June 17	15	05	June 25	04	26
July 1	19	04	July 9	02	40	July 17	04	40	July 24	09	41
July 31	03	53	Aug. 7	19	27	Aug. 15	16	38	Aug. 22	14	17
Aug. 29	14	45	Sept. 6	13	27	Sept. 14	03	10	Sept. 20	19	48
Sept. 28	04	08	Oct. 6	07	46	Oct. 13	12	51	Oct. 20	03	41
Oct. 27	20	14	Nov. 5	01	09	Nov. 11	22	28	Nov. 18	14	54
Nov. 26	14	39	Dec. 4	16	23	Dec. 11	08	42	Dec. 18	05	49
Dec. 26	10	11									

1982

			Jan. 3	04	47	Jan. 9	19	54	Jan. 16	23	59
Jan. 25	04	57	Feb. 1	14	29	Feb. 8	07	58	Feb. 15	20	22
Feb. 23	21	14	Mar. 2	22	16	Mar. 9	20	46	Mar. 17	17	15
Mar. 25	10	18	Apr. 1	05	09	Apr. 8	10	19	Apr. 16	12	43
Apr. 23	20	29	Apr. 30	12	08	May 8	00	45	May 16	05	12
May 23	04	41	May 29	20	07	June 6	16	00	June 14	18	07
June 21	11	53	June 28	05	58	July 6	07	32	July 14	03	47
July 20	18	58	July 27	18	22	Aug. 4	22	35	Aug. 12	11	09
Aug. 19	02	46	Aug. 26	09	50	Sept. 3	12	29	Sept. 10	17	20
Sept. 17	12	10	Sept. 25	04	07	Oct. 3	01	10	Oct. 9	23	28
Oct. 17	00	05	Oct. 25	00	09	Nov. 1	12	58	Nov. 8	06	39
Nov. 15	15	11	Nov. 23	20	06	Dec. 1	00	22	Dec. 7	15	54
Dec. 15	09	19	Dec. 23	14	18	Dec. 30	11	34			

PHASES OF THE MOON

NEW MOON				FIRST QUARTER				FULL MOON				LAST QUARTER			
	D	H	M		D	H	M		D	H	M		D	H	M
1983															
												Jan.	6	04	01
Jan.	14	05	09	Jan.	22	05	35	Jan.	28	22	27	Feb.	4	19	18
Feb.	13	00	32	Feb.	20	17	33	Feb.	27	08	59	Mar.	6	13	17
Mar.	14	17	45	Mar.	22	02	27	Mar.	28	19	28	Apr.	5	08	40
Apr.	13	07	59	Apr.	20	08	58	Apr.	27	06	32	May	5	03	44
May	12	19	26	May	19	14	18	May	26	18	48	June	3	21	08
June	11	04	38	June	17	19	47	June	25	08	33	July	3	12	13
July	10	12	19	July	17	02	52	July	24	23	28	Aug.	2	00	53
Aug.	8	19	19	Aug.	15	12	48	Aug.	23	15	00	Aug.	31	11	23
Sept.	7	02	36	Sept.	14	02	25	Sept.	22	06	37	Sept.	29	20	06
Oct.	6	11	17	Oct.	13	19	43	Oct.	21	21	54	Oct.	29	03	38
Nov.	4	22	22	Nov.	12	15	49	Nov.	20	12	30	Nov.	27	10	51
Dec.	4	12	27	Dec.	12	13	10	Dec.	20	02	02	Dec.	26	18	53
1984															
Jan.	3	05	16	Jan.	11	09	49	Jan.	18	14	06	Jan.	25	04	48
Feb.	1	23	47	Feb.	10	04	00	Feb.	17	00	42	Feb.	23	17	13
Mar.	2	18	32	Mar.	10	18	28	Mar.	17	10	10	Mar.	24	08	00
Apr.	1	12	11	Apr.	9	04	53	Apr.	15	19	11	Apr.	23	00	27
May	1	03	46	May	8	11	50	May	15	04	29	May	22	17	45
May	30	16	49	June	6	16	42	June	13	14	42	June	21	11	11
June	29	03	20	July	5	21	05	July	13	02	21	July	21	04	02
July	28	11	52	Aug.	4	02	34	Aug.	11	15	45	Aug.	19	19	42
Aug.	26	19	26	Sept.	2	10	30	Sept.	10	07	02	Sept.	18	09	32
Sept.	25	03	12	Oct.	1	21	53	Oct.	9	23	59	Oct.	17	21	14
Oct.	24	12	09	Oct.	31	13	09	Nov.	8	17	43	Nov.	16	07	00
Nov.	22	22	58	Nov.	30	08	01	Dec.	8	10	54	Dec.	15	15	26
Dec.	22	11	48	Dec.	30	05	28								

PHASES OF THE MOON

NEW MOON D H M	FIRST QUARTER D H M	FULL MOON D H M	LAST QUARTER D H M
1985			
		Jan. 7 02 18	Jan. 13 23 27
Jan. 21 02 30	Jan. 29 03 31	Feb. 5 15 19	Feb. 12 07 57
Feb. 19 18 44	Feb. 27 23 42	Mar. 7 02 14	Mar. 13 17 35
Mar. 21 11 59	Mar. 29 16 12	Apr. 5 11 33	Apr. 12 04 42
Apr. 20 05 22	Apr. 28 04 26	May 4 19 54	May 11 17 35
May 19 21 41	May 27 12 56	June 3 03 51	June 10 08 20
June 18 11 59	June 25 18 54	July 2 12 09	July 10 00 50
July 17 23 58	July 24 23 40	July 31 21 42	Aug. 8 18 29
Aug. 16 10 07	Aug. 23 04 38	Aug. 30 09 28	Sept. 7 12 17
Sept. 14 19 21	Sept. 21 11 04	Sept. 29 00 09	Oct. 7 05 05
Oct. 14 04 34	Oct. 20 20 13	Oct. 28 17 38	Nov. 5 20 08
Nov. 12 14 21	Nov. 19 09 04	Nov. 27 12 43	Dec. 5 09 02
Dec. 12 00 55	Dec. 19 01 59	Dec. 27 07 31	
1986			
			Jan. 3 19 49
Jan. 10 12 23	Jan. 17 22 14	Jan. 26 00 31	Feb. 2 04 42
Feb. 9 00 57	Feb. 16 19 56	Feb. 24 15 03	Mar. 3 12 18
Mar. 10 14 52	Mar. 18 16 39	Mar. 26 03 03	Apr. 1 19 31
Apr. 9 06 09	Apr. 17 10 35	Apr. 24 12 48	May 1 03 23
May 8 22 10	May 17 01 01	May 23 20 46	May 30 12 56
June 7 14 01	June 15 12 01	June 22 03 42	June 29 00 54
July 7 04 56	July 14 20 11	July 21 10 41	July 28 15 35
Aug. 5 18 36	Aug. 13 02 22	Aug. 19 18 55	Aug. 27 08 39
Sept. 4 07 11	Sept. 11 07 42	Sept. 18 05 35	Sept. 26 03 18
Oct. 3 18 55	Oct. 10 13 30	Oct. 17 19 23	Oct. 25 22 27
Nov. 2 06 03	Nov. 8 21 12	Nov. 16 12 13	Nov. 24 16 51
Dec. 1 16 44	Dec. 8 08 03	Dec. 16 07 06	Dec. 24 09 18
Dec. 31 03 11			

PHASES OF THE MOON

NEW MOON			FIRST QUARTER			FULL MOON			LAST QUARTER		
D	H	M	D	H	M	D	H	M	D	H	M

1987

			Jan. 6	22	36	Jan. 15	02	31	Jan. 22	22	47
Jan. 29	13	46	Feb. 5	16	21	Feb. 13	20	59	Feb. 21	08	57
Feb. 28	00	51	Mar. 7	11	58	Mar. 15	13	13	Mar. 22	16	22
Mar. 29	12	47	Apr. 6	07	49	Apr. 14	02	32	Apr. 20	22	16
Apr. 28	01	36	May 6	02	26	May 13	12	51	May 20	04	04
May 27	15	14	June 4	18	54	June 11	20	50	June 18	11	03
June 26	05	37	July 4	08	35	July 11	03	33	July 17	20	18
July 25	20	38	Aug. 2	19	25	Aug. 9	10	18	Aug. 16	08	26
Aug. 24	12	00	Sept. 1	03	49	Sept. 7	18	14	Sept. 14	23	45
Sept. 23	03	09	Sept. 30	10	40	Oct. 7	04	13	Oct. 14	18	06
Oct. 22	17	29	Oct. 29	17	11	Nov. 5	16	47	Nov. 13	14	39
Nov. 21	06	34	Nov. 28	00	38	Dec. 5	08	02	Dec. 13	11	42
Dec. 20	18	26	Dec. 27	10	01						

1988

						Jan. 4	01	41	Jan. 12	07	04
Jan. 19	05	27	Jan. 25	21	55	Feb. 2	20	53	Feb. 10	23	01
Feb. 17	15	56	Feb. 24	12	16	Mar. 3	16	02	Mar. 11	10	57
Mar. 18	02	03	Mar. 25	04	42	Apr. 2	09	22	Apr. 9	19	22
Apr. 16	12	01	Apr. 23	22	32	May 1	23	42	May 9	01	24
May 15	22	11	May 23	16	50	May 31	10	54	June 7	06	23
June 14	09	15	June 22	10	24	June 29	19	47	July 6	11	38
July 13	21	53	July 22	02	16	July 29	03	26	Aug. 4	18	23
Aug. 12	12	32	Aug. 20	15	52	Aug. 27	10	57	Sept. 3	03	52
Sept. 11	04	51	Sept. 19	03	19	Sept. 25	19	08	Oct. 2	16	59
Oct. 10	21	50	Oct. 18	13	02	Oct. 25	04	36	Nov. 1	10	12
Nov. 9	14	21	Nov. 16	21	36	Nov. 23	15	54	Dec. 1	06	50
Dec. 9	05	37	Dec. 16	05	41	Dec. 23	05	29	Dec. 31	04	58

PHASES OF THE MOON

1989

NEW MOON			FIRST QUARTER			FULL MOON			LAST QUARTER		
	D	H M		D	H M		D	H M		D	H M
Jan.	7	19 23	Jan.	14	14 00	Jan.	21	21 34	Jan.	30	02 04
Feb.	6	07 38	Feb.	12	23 15	Feb.	20	15 33	Feb.	28	20 08
Mar.	7	18 20	Mar.	14	10 11	Mar.	22	09 59	Mar.	30	10 23
Apr.	6	03 34	Apr.	12	23 14	Apr.	21	03 14	Apr.	28	20 46
May	5	11 48	May	12	14 21	May	20	18 17	May	28	04 01
June	3	19 54	June	11	07 00	June	19	06 58	June	26	09 10
July	3	05 00	July	11	00 20	July	18	17 43	July	25	13 33
Aug.	1	16 06	Aug.	9	17 29	Aug.	17	03 07	Aug.	23	18 41
Aug.	31	05 45	Sept.	8	09 50	Sept.	15	11 51	Sept.	22	02 11
Sept.	29	21 48	Oct.	8	00 53	Oct.	14	20 33	Oct.	21	13 19
Oct.	29	15 29	Nov.	6	14 12	Nov.	13	05 52	Nov.	20	04 45
Nov.	28	09 42	Dec.	6	01 26	Dec.	12	16 31	Dec.	19	23 55
Dec.	28	03 21									

1990

NEW MOON			FIRST QUARTER			FULL MOON			LAST QUARTER		
	D	H M		D	H M		D	H M		D	H M
			Jan.	4	10 41	Jan.	11	04 58	Jan.	18	21 18
Jan.	26	19 21	Feb.	2	18 33	Feb.	9	19 17	Feb.	17	18 49
Feb.	25	08 55	Mar.	4	02 06	Mar.	11	10 59	Mar.	19	14 31
Mar.	26	19 49	Apr.	2	10 25	Apr.	10	03 20	Apr.	18	07 03
Apr.	25	04 28	May	1	20 19	May	9	19 31	May	17	19 46
May	24	11 48	May	31	08 12	June	8	11 02	June	16	04 49
June	22	18 56	June	29	22 08	July	8	01 24	July	15	11 05
July	22	02 55	July	29	14 02	Aug.	6	14 21	Aug.	13	15 55
Aug.	20	12 40	Aug.	28	07 36	Sept.	5	01 47	Sept.	11	20 54
Sept.	19	00 47	Sept.	27	02 07	Oct.	4	12 03	Oct.	11	03 32
Oct.	18	15 38	Oct.	26	20 27	Nov.	2	21 49	Nov.	9	13 02
Nov.	17	09 05	Nov.	25	13 12	Dec.	2	07 51	Dec.	9	02 05
Dec.	17	04 23	Dec.	25	03 17	Dec.	31	18 37			

PHASES OF THE MOON

NEW MOON			FIRST QUARTER			FULL MOON			LAST QUARTER		
D	H	M	D	H	M	D	H	M	D	H	M

1991

									Jan.	7	18 37
Jan.	15	23 51	Jan.	23	14 23	Jan.	30	06 10	Feb.	6	13 53
Feb.	14	17 33	Feb.	21	22 59	Feb.	28	18 26	Mar.	8	10 33
Mar.	16	08 11	Mar.	23	06 03	Mar.	30	07 18	Apr.	7	06 47
Apr.	14	19 38	Apr.	21	12 40	Apr.	28	21 00	May	7	00 48
May	14	04 37	May	20	19 47	May	28	11 38	June	5	15 31
June	12	12 07	June	19	04 20	June	27	03 00	July	5	02 51
July	11	19 07	July	18	15 12	July	26	18 25	Aug.	3	11 27
Aug.	10	02 28	Aug.	17	05 02	Aug.	25	09 08	Sept.	1	18 17
Sept.	8	11 02	Sept.	15	22 02	Sept.	23	22 41	Oct.	1	00 31
Oct.	7	21 39	Oct.	15	17 34	Oct.	23	11 09	Oct.	30	07 12
Nov.	6	11 12	Nov.	14	14 02	Nov.	21	22 58	Nov.	28	15 22
Dec.	6	03 57	Dec.	14	09 33	Dec.	21	10 24	Dec.	28	01 56

1992

Jan.	4	23 11	Jan.	13	02 33	Jan.	19	21 30	Jan.	26	15 28
Feb.	3	19 00	Feb.	11	16 16	Feb.	18	08 04	Feb.	25	07 57
Mar.	4	13 24	Mar.	12	02 37	Mar.	18	18 19	Mar.	26	02 31
Apr.	3	05 02	Apr.	10	10 07	Apr.	17	04 44	Apr.	24	21 40
May	2	17 46	May	9	15 45	May	16	16 04	May	24	15 54
June	1	03 57	June	7	20 48	June	15	04 51	June	23	08 12
June	30	12 19	July	7	02 45	July	14	19 07	July	22	22 13
July	29	19 36	Aug.	5	10 59	Aug.	13	10 28	Aug.	21	10 02
Aug.	28	02 43	Sept.	3	22 40	Sept.	12	02 17	Sept.	19	19 54
Sept.	26	10 41	Oct.	3	14 13	Oct.	11	18 04	Oct.	19	04 13
Oct.	25	20 35	Nov.	2	09 12	Nov.	10	09 21	Nov.	17	11 40
Nov.	24	09 12	Dec.	2	06 18	Dec.	9	23 41	Dec.	16	19 15
Dec.	24	00 43									

PHASES OF THE MOON

	NEW MOON			FIRST QUARTER			FULL MOON			LAST QUARTER		
	D	H	M	D	H	M	D	H	M	D	H	M

1993

	NEW MOON			FIRST QUARTER			FULL MOON			LAST QUARTER		
				Jan. 1	03	39	Jan. 8	12	38	Jan. 15	04	02
Jan. 22	18	28	Jan. 30	23	20	Feb. 6	23	56	Feb. 13	14	58
Feb. 21	13	06	Mar. 1	15	47	Mar. 8	09	47	Mar. 15	04	17
Mar. 23	07	16	Mar. 31	04	11	Apr. 6	18	45	Apr. 13	19	39
Apr. 21	23	49	Apr. 29	12	42	May 6	03	35	May 13	12	20
May 21	14	08	May 28	18	22	June 4	13	03	June 12	05	37
June 20	01	54	June 26	22	45	July 3	23	46	July 11	22	50
July 19	11	25	July 26	03	26	Aug. 2	12	11	Aug. 10	15	20
Aug. 17	19	29	Aug. 24	09	59	Sept. 1	02	34	Sept. 9	06	27
Sept. 16	03	12	Sept. 22	19	33	Sept. 30	18	54	Oct. 8	19	37
Oct. 15	11	37	Oct. 22	08	53	Oct. 30	12	38	Nov. 7	06	37
Nov. 13	21	35	Nov. 21	02	04	Nov. 29	06	32	Dec. 6	15	50
Dec. 13	09	28	Dec. 20	22	26	Dec. 28	23	07			

1994

	NEW MOON			FIRST QUARTER			FULL MOON			LAST QUARTER		
										Jan. 5	00	01
Jan. 11	23	11	Jan. 19	20	27	Jan. 27	13	24	Feb. 3	08	07
Feb. 10	14	31	Feb. 18	17	49	Feb. 26	01	16	Mar. 4	16	55
Mar. 12	07	06	Mar. 20	12	15	Mar. 27	11	11	Apr. 3	02	55
Apr. 11	00	18	Apr. 19	02	35	Apr. 25	19	46	May 2	14	33
May 10	17	08	May 18	12	51	May 25	03	40	June 1	04	03
June 9	08	28	June 16	19	57	June 23	11	34	June 30	19	31
July 8	21	39	July 16	01	13	July 22	20	17	July 30	12	41
Aug. 7	08	46	Aug. 14	05	58	Aug. 21	06	48	Aug. 29	06	41
Sept. 5	18	34	Sept. 12	11	35	Sept. 19	20	02	Sept. 28	00	24
Oct. 5	03	56	Oct. 11	19	18	Oct. 19	12	19	Oct. 27	16	45
Nov. 3	13	37	Nov. 10	06	15	Nov. 18	06	58	Nov. 26	07	05
Dec. 2	23	55	Dec. 9	21	08	Dec. 18	02	18	Dec. 25	19	08

PHASES OF THE MOON

	NEW MOON				FIRST QUARTER				FULL MOON				LAST QUARTER		
	D	H	M		D	H	M		D	H	M		D	H	M
1995															
Jan.	1	10	57	Jan.	8	15	47	Jan.	16	20	27	Jan.	24	04	59
Jan.	30	22	48	Feb.	7	12	55	Feb.	15	12	16	Feb.	22	13	05
Mar.	1	11	49	Mar.	9	10	15	Mar.	17	01	26	Mar.	23	20	11
Mar.	31	02	09	Apr.	8	05	36	Apr.	15	12	09	Apr.	22	03	19
Apr.	29	17	37	May	7	21	45	May	14	20	49	May	21	11	36
May	29	09	28	June	6	10	27	June	13	04	04	June	19	22	02
June	28	00	51	July	5	20	04	July	12	10	50	July	19	11	11
July	27	15	15	Aug.	4	03	17	Aug.	10	18	17	Aug.	18	03	04
Aug.	26	04	33	Sept.	2	09	04	Sept.	9	03	38	Sept.	16	21	10
Sept.	24	16	56	Oct.	1	14	36	Oct.	8	15	53	Oct.	16	16	27
Oct.	24	04	37	Oct.	30	21	18	Nov.	7	07	21	Nov.	15	11	41
Nov.	22	15	44	Nov.	29	06	29	Dec.	7	01	28	Dec.	15	05	33
Dec.	22	02	24	Dec.	28	19	07								
1996															
								Jan.	5	20	52	Jan.	13	20	47
Jan.	20	12	51	Jan.	27	11	15	Feb.	4	15	59	Feb.	12	08	38
Feb.	18	23	31	Feb.	26	05	53	Mar.	5	09	24	Mar.	12	17	16
Mar.	19	10	46	Mar.	27	01	32	Apr.	4	00	08	Apr.	10	23	37
Apr.	17	22	50	Apr.	25	20	42	May	3	11	49	May	10	05	05
May	17	11	47	May	25	14	15	June	1	20	48	June	8	11	06
June	16	01	37	June	24	05	25	July	1	03	59	July	7	18	56
July	15	16	16	July	23	17	50	July	30	10	37	Aug.	6	05	26
Aug.	14	07	35	Aug.	22	03	38	Aug.	28	17	53	Sept.	4	19	07
Sept.	12	23	09	Sept.	20	11	24	Sept.	27	02	52	Oct.	4	12	06
Oct.	12	14	16	Oct.	19	18	10	Oct.	26	14	12	Nov.	3	07	52
Nov.	11	04	17	Nov.	18	01	10	Nov.	25	04	11	Dec.	3	05	07
Dec.	10	16	58	Dec.	17	09	31	Dec.	24	20	41				

PHASES OF THE MOON

NEW MOON			FIRST QUARTER			FULL MOON			LAST QUARTER		
D	H	M	D	H	M	D	H	M	D	H	M

1997

NEW MOON			FIRST QUARTER			FULL MOON			LAST QUARTER		
									Jan.	2 01	46
Jan.	9 04	26	Jan.	15 20	03	Jan.	23 15	12	Jan.	31 19	41
Feb.	7 15	08	Feb.	14 08	58	Feb.	22 10	28	Mar.	2 09	39
Mar.	9 01	15	Mar.	16 00	07	Mar.	24 04	46	Mar.	31 19	39
Apr.	7 11	03	Apr.	14 17	01	Apr.	22 20	35	Apr.	30 02	37
May	6 20	48	May	14 10	56	May	22 09	14	May	29 07	52
June	5 07	05	June	13 04	53	June	20 19	10	June	27 12	43
July	4 18	41	July	12 21	45	July	20 03	21	July	26 18	30
Aug.	3 08	15	Aug.	11 12	43	Aug.	18 10	57	Aug.	25 02	24
Sept.	1 23	53	Sept.	10 01	33	Sept.	16 18	52	Sept.	23 13	37
Oct.	1 16	53	Oct.	9 12	23	Oct.	16 03	47	Oct.	23 04	50
Oct.	31 10	02	Nov.	7 21	44	Nov.	14 14	13	Nov.	21 23	59
Nov.	30 02	15	Dec.	7 06	11	Dec.	14 02	38	Dec.	21 21	44
Dec.	29 16	58									

1998

NEW MOON			FIRST QUARTER			FULL MOON			LAST QUARTER		
			Jan.	5 14	20	Jan.	12 17	25	Jan.	20 19	41
Jan.	28 06	02	Feb.	3 22	54	Feb.	11 10	24	Feb.	19 15	28
Feb.	26 17	27	Mar.	5 08	42	Mar.	13 04	35	Mar.	21 07	38
Mar.	28 03	15	Apr.	3 20	20	Apr.	11 22	24	Apr.	19 19	54
Apr.	26 11	43	May	3 10	05	May	11 14	30	May	19 04	36
May	25 19	34	June	2 01	46	June	10 04	19	June	17 10	40
June	24 03	52	July	1 18	43	July	9 16	02	July	16 15	15
July	23 13	45	July	31 12	06	Aug.	8 02	11	Aug.	14 19	49
Aug.	22 02	04	Aug.	30 05	08	Sept.	6 11	23	Sept.	13 01	59
Sept.	20 17	02	Sept.	28 21	12	Oct.	5 20	13	Oct.	12 11	12
Oct.	20 10	10	Oct.	28 11	47	Nov.	4 05	20	Nov.	11 00	29
Nov.	19 04	28	Nov.	27 00	24	Dec.	3 15	21	Dec.	10 17	55
Dec.	18 22	44	Dec.	26 10	47						

PHASES OF THE MOON

NEW MOON			FIRST QUARTER			FULL MOON			LAST QUARTER		
D	H	M	D	H	M	D	H	M	D	H	M

1999

						Jan.	2	02 51	Jan.	9	14 23
Jan.	17	15 47	Jan.	24	19 17	Jan.	31	16 08	Feb.	8	11 59
Feb.	16	06 40	Feb.	23	02 44	Mar.	2	07 00	Mar.	10	08 42
Mar.	17	18 49	Mar.	24	10 19	Mar.	31	22 50	Apr.	9	02 52
Apr.	16	04 23	Apr.	22	19 03	Apr.	30	14 56	May	8	17 29
May	15	12 06	May	22	05 35	May	30	06 41	June	7	04 21
June	13	19 04	June	20	18 14	June	28	21 38	July	6	11 58
July	13	02 25	July	20	09 02	July	28	11 26	Aug.	4	17 28
Aug.	11	11 09	Aug.	19	01 48	Aug.	26	23 49	Sept.	2	22 18
Sept.	9	22 03	Sept.	17	20 07	Sept.	25	10 52	Oct.	2	04 03
Oct.	9	11 36	Oct.	17	15 00	Oct.	24	21 04	Oct.	31	12 05
Nov.	8	03 54	Nov.	16	09 04	Nov.	23	07 05	Nov.	29	23 19
Dec.	7	22 32	Dec.	16	00 51	Dec.	22	17 33	Dec.	29	14 05

2000

Jan.	6	18 14	Jan.	14	13 35	Jan.	21	04 41	Jan.	28	07 58
Feb.	5	13 04	Feb.	12	23 22	Feb.	19	16 28	Feb.	27	03 55
Mar.	6	05 18	Mar.	13	07 00	Mar.	20	04 45	Mar.	28	00 22
Apr.	4	18 14	Apr.	11	13 31	Apr.	18	17 43	Apr.	26	19 31
May	4	04 13	May	10	20 01	May	18	07 36	May	26	11 56
June	2	12 16	June	9	03 31	June	16	22 28	June	25	01 01
July	1	19 21	July	8	12 55	July	16	13 56	July	24	11 03
July	31	02 26	Aug.	7	01 03	Aug.	15	05 14	Aug.	22	18 52
Aug.	29	10 20	Sept.	5	16 28	Sept.	13	19 38	Sept.	21	01 29
Sept.	27	19 54	Oct.	5	11 00	Oct.	13	08 54	Oct.	20	08 00
Oct.	27	07 59	Nov.	4	07 27	Nov.	11	21 15	Nov.	18	15 26
Nov.	25	23 13	Dec.	4	03 56	Dec.	11	09 04	Dec.	18	00 43
Dec.	25	17 22									

PHASES OF THE MOON

	NEW MOON			FIRST QUARTER				FULL MOON				LAST QUARTER			
	D	H	M		D	H	M		D	H	M		D	H	M

2001

	NEW MOON			FIRST QUARTER				FULL MOON				LAST QUARTER			
				Jan.	2	22	32	Jan.	9	20	26	Jan.	16	12	36
Jan.	24	13	08	Feb.	1	14	04	Feb.	8	07	13	Feb.	15	03	25
Feb.	23	08	22	Mar.	3	02	04	Mar.	9	17	24	Mar.	16	20	46
Mar.	25	01	22	Apr.	1	10	50	Apr.	8	03	23	Apr.	15	15	32
Apr.	23	15	27	Apr.	30	17	09	May	7	13	54	May	15	10	11
May	23	02	47	May	29	22	10	June	6	01	40	June	14	03	29
June	21	11	59	June	28	03	21	July	5	15	05	July	13	18	46
July	20	19	46	July	27	10	09	Aug.	4	05	57	Aug.	12	07	54
Aug.	19	02	57	Aug.	25	19	56	Sept.	2	21	44	Sept.	10	19	00
Sept.	17	10	28	Sept.	24	09	32	Oct.	2	13	49	Oct.	10	04	21
Oct.	16	19	24	Oct.	24	02	59	Nov.	1	05	42	Nov.	8	12	22
Nov.	15	06	41	Nov.	22	23	21	Nov.	30	20	50	Dec.	7	19	53
Dec.	14	20	48	Dec.	22	20	57	Dec.	30	10	42				

2002

	NEW MOON			FIRST QUARTER				FULL MOON				LAST QUARTER			
												Jan.	6	03	56
Jan.	13	13	30	Jan.	21	17	48	Jan.	28	22	52	Feb.	4	13	34
Feb.	12	07	41	Feb.	20	12	03	Feb.	27	09	18	Mar.	6	01	26
Mar.	14	02	04	Mar.	22	02	30	Mar.	28	18	26	Apr.	4	15	30
Apr.	12	19	22	Apr.	20	12	49	Apr.	27	03	01	May	4	07	17
May	12	10	47	May	19	19	43	May	26	11	53	June	3	00	06
June	10	23	47	June	18	00	31	June	24	21	43	July	2	17	21
July	10	10	27	July	17	04	48	July	24	09	08	Aug.	1	10	24
Aug.	8	19	16	Aug.	15	10	13	Aug.	22	22	30	Aug.	31	02	32
Sept.	7	03	11	Sept.	13	18	10	Sept.	21	14	00	Sept.	29	17	04
Oct.	6	11	19	Oct.	13	05	35	Oct.	21	07	21	Oct.	29	05	29
Nov.	4	20	36	Nov.	11	20	54	Nov.	20	01	35	Nov.	27	15	47
Dec.	4	07	35	Dec.	11	15	49	Dec.	19	19	11	Dec.	27	00	32

PHASES OF THE MOON

NEW MOON			FIRST QUARTER			FULL MOON			LAST QUARTER		
D	H	M	D	H	M	D	H	M	D	H	M

2003

NEW MOON			FIRST QUARTER			FULL MOON			LAST QUARTER		
Jan.	2 20 24		Jan.	10 13 16		Jan.	18 10 49		Jan.	25 08 35	
Feb.	1 10 50		Feb.	9 11 13		Feb.	16 23 52		Feb.	23 16 47	
Mar.	3 02 36		Mar.	11 07 17		Mar.	18 10 35		Mar.	25 01 53	
Apr.	1 19 19		Apr.	9 23 41		Apr.	16 19 37		Apr.	23 12 20	
May	1 12 15		May	9 11 54		May	16 03 37		May	23 00 31	
May	31 04 20		June	7 20 28		June	14 11 17		June	21 14 46	
June	29 18 40		July	7 02 33		July	13 19 22		July	21 07 02	
July	29 06 54		Aug.	5 07 29		Aug.	12 04 50		Aug.	20 00 49	
Aug.	27 17 27		Sept.	3 12 36		Sept.	10 16 38		Sept.	18 19 04	
Sept.	26 03 10		Oct.	2 19 11		Oct.	10 07 28		Oct.	18 12 32	
Oct.	25 12 51		Nov.	1 04 26		Nov.	9 01 14		Nov.	17 04 17	
Nov.	23 23 01		Nov.	30 17 17		Dec.	8 20 37		Dec.	16 17 43	
Dec.	23 09 45		Dec.	30 10 04							

Bibliography

Adderly, E. E., and Bowen, E. G., "Lunar Component in Precipitation Data," *Science* 137 (1962):749–50.

American Ephemeris and Nautical Almanac (Washington, D.C.: U. S. Government Printing Office, Annual editions, 1956–70).

Andrews, E. A., "Moon Talk: The Cyclic Periodicity of Postoperative Hemorrhage," *Journal of the Florida Medical Association* 46 (1961):1362–66.

Armstrong, N., Collins, M., and Aldrin, E. E., Jr., epilogue by Arthur C. Clarke, *First on the Moon* (Boston: Little, Brown & Co., 1970).

Barnett, L., *The Universe and Dr. Einstein* (New York: William Sloane Associates, 1948).

Bauer, S. F., and Hornick, E. J., "Lunar Effect on Mental Illness," *American Journal of Psychiatry* 125 (1968):696–97.

Becker, R. O., "Electromagnetic Forces and Life Processes," *Technology Review* 75 (1972):2–8.

—— Letter to Jerome Agel, November 24, 1975.

Bisbee, D. S. P., "The Effects of Lunar Cycles and Diurnal Rhythms on Activity, Exploration and Elicited Aggression in Rats and Mice," Thesis for Ph.D. in experimental psychology, Auburn University, 1974.

Bjerknes, J., "Atmospheric Tides," *Journal of Marine Research* 7 (1948):157–62.

Blackman, S., and Catalina, D., "The Moon and the Emergency Room," *Perceptual and Motor Skills* 37 (1973):624–26.

Bokonjic, R., and Zec, N., "Strokes and the Weather: A Quantitative Statistical Study," *Journal of Neurological Science* 6 (1968):483–91.

Bradley, D. A., Woodbury, M. A., and Brier, G. W., "Lunar Synodical Period and Widespread Precipitation," *Science* 137 (1962):748–49.

Brier, G. W., "Diurnal and Semidiurnal Atmospheric Tides in

Relation to Precipitation Variations," *Monthly Weather Review* 93 (1965):93–100.

—— and Bradley, D. A., "The Lunar Synodical Period and Precipitation in the United States," *Journal of Atmospheric Science* 21 (1964):386–95.

—— and Simpson, J., "Tropical Cloudiness and Rainfall Related to Tidal Pressure and Variations," *Quarterly Journal of the Royal Meteorological Society* 95 (1969):120–47.

Brown, F. A., "Biological Clocks: Endogenous Cycles Synchronized by Subtle Geophysical Rhythms," *Biosystems* 8 (1976):67–81.

—— "The 'Clocks' Timing Biological Rhythms," *American Scientist* 60 (1972):756–66.

—— and Chow, C. S., "Field Interactions Between Organisms," *Biological Bulletin* 145 (1973):437–61.

—— "Lunar-Correlated Variations in Water Uptake by Bean Seeds," *Biological Bulletin* 145 (1973):265–78.

—— Hastings, J. W., and Palmer, J. D., *The Biological Clock: Two Views* (New York: Academic Press, 1970).

Bulka, Rabbi R. P., Letter to Dr. Arnold L. Lieber, August 6, 1972.

Burke, P., Letter to Dr. Arnold L. Lieber, December 12, 1972.

Burr, H. S., *Blueprint for Immortality* (London: Neville Spearman, 1972).

Calder, N., *Violent Universe* (New York: The Viking Press, 1969).

Carpenter, T. H., Holle, R. L., and Fernandez-Partagas, J. J., "Observed Relationships Between Lunar Tidal Cycles and Formation of Hurricanes and Tropical Storms," *Monthly Weather Review* 100 (1972):451–60.

Chancellor, J., *Charles Darwin* (New York: Taplinger Pub. Co., 1976).

Chapman, L. J., "A Search for Lunacy," *Journal of Nervous and Mental Disease* 132 (1961):171–74.

Cooper, H. S. F., Jr., *Moon Rocks* (New York: The Dial Press, 1970).

Cutts, B., "The Marital Blitz Bliss," The Atlanta *Constitution*, September 2, 1977.

Danneel, R., "The Influence of Geophysical Factors on the Frequency of Suicides" [in German], *Archive für Psychiatrie und Nervenkrankenheiten* 219 (1974):153–57.

Darwin, C., *The Descent of Man and Selection in Relation to Sex* (New York: AMS Press, 1972).

Davidson, J. E., and others, "Intriguing Accident Patterns Plotted Against a Background of Natural Environment Features," Sandia Laboratories, New Mexico, August 1970.

Decimal Ephemeris of the Sun and Moon, 1848 to 1974 (Washington, D.C.: U. S. Weather Bureau microfilm.)

de Rudder, B., *About the So-called Cosmic Rhythms in Humans* [in German], (Stuttgart: G. Thierne, 1948).

Deshpanday, S. K., Letter to Dr. Arnold L. Lieber and Dr. Carolyn R. Sherin, November 18, 1972.

Dewan, E. J., "On the Possibility of a Perfect Rhythm Method of Birth Control by Periodic Light Stimulation," *American Journal of Obstetrics and Gynecology* 98 (1967):656–59.

Dorland, J. A., Letter to Dr. Arnold L. Lieber, September 9, 1972.

Dubrov, A., "Heliobiology," *Soviet Life*, January 1972.

Edland, J. F., "Suicide by Automobile," *Albany Law Review* 36 (1972):536–42.

Eisler, R., *Man Into Wolf, An Anthropological Interpretation of Sadism, Masochism and Lycanthropy* (New York: Greenwood Press, 1969).

Ferenczi, S., *Thalassa: A Theory of Genitality* (New York: W. W. Norton & Co., 1968).

Florence, E., Letter to Dr. Arnold L. Lieber, August 17, 1972.

Fox, H. M., "Lunar Periodicity of Reproduction," *Nature* 130 (1932):23.

Fox, V., Letter to Jerome Agel, January 21, 1977.

Friedman, H., and Becker, R. O., "Geomagnetic Parameters and Psychiatric Hospital Admissions," *Nature* 200 (1963):626–28.

—— Becker, R. O., and Bachman, C. H., "Effect of Magnetic Fields on Reaction Time Performance," *Nature* 213 (1967):949–56.

—— "Psychiatric Ward Behavior and Geophysical Parameters," *Nature* 205 (1965):1050–52.

Gauquelin, M., *The Cosmic Clocks* (Chicago: Henry Regnery Co., 1967).

Geller, S. H., and Shannon, H., "The Moon, Weather and Mental Hospital Contacts: Confirmation and Explanation of the Transylvania Effect," *Journal of Psychiatric Nursing and Mental Health Services* 14 (1976):13–17.

Graves, R., *The Greek Myths*, Vol. 1 (Baltimore, Md.: Penguin Books, 1955).

—— *The White Goddess* (New York: Octagon, 1972).

Gribbin, J. R., and Plagemann, S. H., *The Jupiter Effect* (New York: Random House, 1974).

Gunn, D. L., and Jenkins, P. M., "Lunar Periodicity in Homo Sapiens," *Nature* 139 (1937):84.

Guthmann, H., and Oswald, D., "Menstruation und Mond," *Manschrift für Geburtsch und Gynekologie* 103 (1936):232–35.

Haddock, J. S., Letter to Dr. Arnold L. Lieber, May 5, 1972.

Harding, J. E., *Woman's Mysteries* (New York: Putnam, 1972).

Heckert, H., *Lunar Rhythms of Human Organisms* (Leipzig: Academic Publishing Company, 1961 [in German]).

—— "Monthly Sexual Periodicity in the Male," Summary, *Chronobiologia*, 11 (1975):Supp. 1.

Inglis, S. J., *Planets, Stars and Galaxies* (New York: John Wiley and Sons, 1967).

Jordan, D. L., "Tidal Forces and the Formation of Hurricanes," *Nature* 175 (1955):38–39.

Jung, C. G., *Synchronicity: An Acausal Connecting Principle*, translated by R. F. C. Hull (Princeton, N.J.: Bollingen Series, Princeton University Press, 1969).

Kelley, D. M., "Mania and the Moon," *Psychoanalytical Review* 29 (1942):406–26.

Kerry, R. J., and Owen, G., "Lithium Carbonate as a Mood and Total Body Water Stabilizer," *Archives of General Psychiatry* 22 (1970):301–3.

Kiser, W. L., Carpenter, T. H., and Brier, G. W., "The Atmospheric Tides at Wake Island," *Monthly Weather Review* 91 (1963):566–72.

Klinowska, M., "A Comparison of the Lunar and Solar Activity Rhythms of the Golden Hamster," *Journal of Interdisciplinary Cycle Research* 3 (1972):145–50.

—— "Lunar Rhythms in Activity, Urinary Volume and Acidity in the Golden Hamster," *Journal of Interdisciplinary Cycle Research* 1 (1970):317–22.

Koestler, A., *The Watershed: A Biography of Johannes Kepler* (Garden City, N.Y.: Anchor Books, 1960).

Kolisko, L., *The Moon and the Growth of Plants* (Bray-on-Thames: Anthroposophical Agricultural Foundation, 1936).

Krippner, S., and Rubin, D., eds., *The Kirlian Aura: Photographing* the Galaxies of Life (Garden City, N.Y.: Anchor Press, 1974).

Lakein, A., *It's About Time & It's About Time,* produced by Jerome Agel (New York: Bantam Books, 1975).

Lester, D., Brockopp, G. W., and Priebe, K., "Association Between Full Moon and Completed Suicide," *Psychological Reports* 25 (1969):598.

Lieber, A. L., "Human Aggression and the Lunar Synodic Cycle," paper presented at the Seventh International Biometeorological Congress, College Park, Md., August 1975.

—— "Lunar Effect on Homicides: A Confirmation," *International Journal of Chronobiology* 4 (1973):338–39.

—— "On the Moon Again," letter to editor, *American Journal of Psychiatry* 132 (1975):669–70.

—— Letter to Dr. Joseph Davis, Dade County Medical Examiner, January 1974.

—— and Sherin, C. R., "Homicides and the Lunar Cycle: Toward a Theory of Lunar Influence on Human Emotional Disturbance," *American Journal of Psychiatry* 129 (1972):69–74.

Lilienfeld, D. M., "Lunar Effect on Mental Illness," *American Journal of Psychiatry* 125 (1969):1454.

Luce, G. G., ed., *Biological Rhythms in Psychiatry and Medicine* (Washington, D.C.: Public Health Service Publication #2088, U. S. Government Printing Office, 1970).

Malmstrom, E. J., "Correlating Crime with Lunar Cycles," National Institute of Mental Health Report, December 1977.

Marks, R. W., ed., *The New Dictionary and Handbook of Aerospace* (New York: Bantam Books, 1969).

McDaniel, W. B., "The Moon, Werewolves, and Medecine," *Transactions and Studies of the College of Physicians of Philadelphia,* 4 Series, 18, 3 (December 1950).

McDonald, R. L., "Lunar and Seasonal Variations in Obstetric Factors," *Journal of Genetic Psychology* 108 (1966):81–87.

Menaker, W., "Lunar Periodicity with Reference to Live Births," *American Journal of Obstetrics and Gynecology* 98 (1967): 1002–4.

—— and Menaker, A., "Lunar Periodicity in Human Reproduction: A Likely Unit of Biological Time," *American Journal of Obstetrics and Gynecology* 77 (1959):905–14.

Miami Herald, "Blame Moon in Triangle?" (June 5, 1976).

Miles, L. E. M., Raynal, D. M., and Wilson, M. A., "Blind Man

Living in Normal Society Has Circadian Rhythms of 24.9
Hours," *Science* 198 (1977):421–23.

Newton, E. C., Letter to Dr. Arnold L. Lieber, May 15, 1972.

Oliven, J. F., "Moonlight and Nervous Disorders," *American Journal of Psychiatry* 99 (1943):579–84.

Osborn, R. D., "The Moon and the Mental Hospital," *Journal of Psychiatric Nursing* 6 (1968):88–93.

Osley, M., Summerville, D., and Borst, L. B., "Natality and the Moon," *American Journal of Obstetrics and Gynecology* 117 (1973):413–15.

Ossenkopp, K.-P., Letter to Dr. Arnold L. Lieber, September 2, 1972.

—— and Ossenkopp, M. D., "Self-inflicted Injuries and the Lunar Cycle: A Preliminary Report," *Journal of Interdisciplinary Cycle Research* 4 (1973):337–48.

Ostrander, S., and Schroeder, L., *Psychic Discoveries Behind the Iron Curtain* (Englewood Cliffs, N.J.: Prentice-Hall, 1970).

Palmer, J. D., *An Introduction to Biological Rhythms* (New York: Harcourt Brace Jovanovich, 1976).

Petersen, W. F., *Man, Weather, Sun* (Springfield, Ill.: Charles C. Thomas, 1947).

Petterssen, S., *Introduction to Meteorology* (New York: McGraw-Hill, 1969).

Phillips, R., *Moonstruck: An Anthology of Lunar Poetry* (New York: The Vanguard Press, 1974).

Pochobradsky, J., "Independence of Human Menstruation on Lunar Phases and Days of the Week," *American Journal of Obstetrics and Gynecology* 118 (1974):1136–38.

Pokorny, A. D., "Moon Phases and Mental Admissions," *Journal of Psychiatric Nursing* 6 (1968):325–27.

—— "Moon Phases, Suicide and Homicide," *American Journal of Psychiatry* 121 (1964):66–67.

—— and Jachimczyk, J., "The Questionable Relationship Between Homicides and the Lunar Cycle," *American Journal of Psychiatry* 131 (1974):827–29.

——, Smith, J. P., and Finch, J. R., "Vehicular Suicides," *Life-threatening Behavior* 2 (1972):105–19.

Presman, A. S., *Electromagnetic Fields and Life* (New York: Plenum Press, 1970).

Ravitz, L. J., "Electromagnetic Field Monitoring of Changing State-Function Including Hypnotic States," *Journal of the*

American Society of Dentistry and Medicine 17 (1970): 119–29.

—— "Electrodynamic Field Theory in Psychiatry," *Southern Medical Journal* 46 (1953):650–60.

—— "Application of Electrodynamic Field Theory in Biology, Psychiatry, Medicine and Hypnosis. I. General Survey," *American Journal of Clinical Hypnosis* 1 (1959):135–50.

—— Letter to Dr. Arnold L. Lieber, July 20, 1972.

Rensberger, B., "186 Top Scientists Dismiss Astrologers as Charlatans," The New York *Times*, September 3, 1975, 1:1.

Rhyne, W. P., "Spontaneous Hemorrhage," *Journal of the Medical Association of Georgia* (December 1966).

Rosenstock, H. A., Vincent, K. R., "A Case of Lycanthropy," *American Journal of Psychiatry* 134 (1977):1147–49.

Rothenberg, A., "The Process of Janusian Thinking in Creativity," *Archives of General Psychiatry* 24 (1971):195–205.

Rounds, H. D., "A Lunar Rhythm in the Occurrence of Blood-Borne Factors in Cockroaches, Mice and Men," *Comparative Biochemical Physiology* (Great Britain: Pergamon Press) 50C, 193–97.

Rush, A. K., *Moon, Moon* (New York: Random House and Berkeley, Ca.: Moon Books, 1976).

Sagan, C., *Other Worlds*, produced by Jerome Agel (New York: Bantam Books, 1975).

Sarton, G., "Lunar Influences on Living Things," *Isis* 30 (1939):498–507.

Selzer, M., and Payne, C., "Automobile Accidents, Suicides and Unconscious Motivation," *American Journal of Psychiatry* 119 (1962):237–40.

Shapiro, J. L., Steiner, D. L., Gray, A. L., and others, "The Moon and Mental Illness," *Perceptual and Motor Skills* 30 (1970):827–30.

"Solar Flares: Link to Thunderstorms," *Science News*, June 18, 1977.

Soyka, F., with Edmonds, A., *The Ion Effect* (New York: E. P. Dutton, 1977).

Stolov, H. L., "Tidal Wind Fields in the Atmosphere," *Journal of Meteorology* 12 (1955):117–40.

Stone, M. H., "Madness and the Moon Revisited," *Psychiatric Annals* 6 (April 1976).

Svens, B., Letter to Jerome Agel, May 2, 1977.

Talbott, S. L., and others, editors of *Pensée. Velikovsky Reconsid-ered* (New York: Warner Books, 1976).

Tasso, J., and Miller, E., "The Effects of the Full Moon on Human Behavior," *Journal of Psychology* 93 (1) (1976): 81–83.

Taylor, L. J., and Diespecker, D. D., "Moon Phases and Suicide Attempts in Australia," *Psychological Reports* 31 (1972):110.

Thomson, D., "The Ebb and Flow of Infection," *Journal of the American Medical Association* 235 (1976):269–72.

Trapp, C. E., "Lunacy and the Moon," *American Journal of Psychiatry* 94 (1937):334–39.

Tromp, S. W., *Medical Biometeorology* (New York: Elsevier, 1963).

—— "Research Project on Terrestrial Biological, Medical and Bio-chemical Phenomena, Caused or Triggered by Possible Extra-terrestrial Physical Forces," *Cycles*, February 1970.

Velikovsky, I., *Worlds in Collision* (New York: Doubleday & Co., 1950).

Verghese, A., and Beig, A., "Public Attitudes Toward Mental Illness—the Vellore Study," *Indian Journal of Psychiatry* (Ma-durai) 16, 1 (1974):8–18.

Vessey, S. H., "Night Observations of Free-Ranging Primates," De-partment of Biology, Bowling Green State University, Ohio.

Walters, E., Markley, R. P., and Tiffany, D. W., "Lunacy: A Type-I Error," *Journal of Abnormal Psychology* 84 (1975):715–17.

Ward, R. R., *The Living Clocks* (New York and Toronto: New American Library, 1971).

Weiskott, G. N., "Moon Phases and Telephone Counseling Calls," *Psychological Reports* 35 (1974):752–54.

—— and Tipton, G. B., "Moon Phases and State Hospital Admis-sions," *Psychological Reports* 37 (1975):486.

Wilson, C., *The Occult* (New York: Random House, 1971).

Wing, L. W., "The Effect of Latitude on Cycles," *Annals of New York Academy of Sciences* 98 (1962).

Wood, F. J., "Predicting Perilous Tides," *Science Year*, The World Book Science Annual, 1975.

Wurtman, R. J., "The Pineal Gland: A Ciba Symposium" (Sum-mary), 1971.

Ziegler, M., "A Lunar Loss of Marbles," San Francisco *Chronicle*, May 2, 1977, 2.

Index